AF812875

Operations Research/Computer Science Interfaces Series

Volume 61

Series Editors:
Ramesh Sharda
Oklahoma State University, Stillwater, Oklahoma, USA

Stefan Voß
University of Hamburg, Hamburg, Germany

More information about this series at http://www.springer.com/series/6375

Marlin Wolf Ulmer

Approximate Dynamic Programming for Dynamic Vehicle Routing

Springer

Marlin Wolf Ulmer
Carl-Friedrich-Gauß-Fakultät
Technische Universität Braunschweig
Braunschweig
Germany

ISSN 1387-666X
Operations Research/Computer Science Interfaces Series
ISBN 978-3-319-55510-2 ISBN 978-3-319-55511-9 (eBook)
DOI 10.1007/978-3-319-55511-9

Library of Congress Control Number: 2017935959

© Springer International Publishing AG 2017
This work is subject to copyright. All rights are reserved by the Publisher, whether the whole or part of the material is concerned, specifically the rights of translation, reprinting, reuse of illustrations, recitation, broadcasting, reproduction on microfilms or in any other physical way, and transmission or information storage and retrieval, electronic adaptation, computer software, or by similar or dissimilar methodology now known or hereafter developed.
The use of general descriptive names, registered names, trademarks, service marks, etc. in this publication does not imply, even in the absence of a specific statement, that such names are exempt from the relevant protective laws and regulations and therefore free for general use.
The publisher, the authors and the editors are safe to assume that the advice and information in this book are believed to be true and accurate at the date of publication. Neither the publisher nor the authors or the editors give a warranty, express or implied, with respect to the material contained herein or for any errors or omissions that may have been made. The publisher remains neutral with regard to jurisdictional claims in published maps and institutional affiliations.

Printed on acid-free paper

This Springer imprint is published by Springer Nature
The registered company is Springer International Publishing AG
The registered company address is: Gewerbestrasse 11, 6330 Cham, Switzerland

The consequences of every act are included in the act itself.

George Orwell, 1984

Foreword

Vehicle routing has received a tremendous attention in recent years. This attention is driven by today's customer expectations with respect to fast and reliable service. On the methodological side the achievements are driven by the remarkable success in overcoming the standard vehicle routing problem towards formulations of "rich" vehicle routing problems. However, the more operational details are incorporated in the "richness" of the problem formulation, the less likely these details will persist in an implementation of the static optimization problem. By accepting the fact that the world is continuously changing, one will also have to accept that a dynamic and stochastic problem formulation does suggest itself.

Typical sources of uncertainty in vehicle routing are stochastic customer requests, stochastic demand of customers as well as stochastic service- and travel times. All these figures may change over time while vehicles are already on the road. Thus, operational planning has to either incorporate possible stochastic changes before the implementation of a plan or subsequent decisions have to be taken while vehicles actually operate. Today's sensor and communication techniques warrant up-to-date mass data for subsequent decision making. Deferring decisions to the latest possible point in time comes along with the highest possible gain of information, but may lose out on advantages to be achieved by taking the right decisions early.

Anticipation of future system states can be seen as a key feature for a successful treatment of dynamic stochastic vehicle routing problems. Thus, in order to take right decisions early, possible impacts have to be defined. The Markov decision process is a suitable instrument for modeling state spaces and transitions within these spaces. Optimization is still present in this modeling approach, but does step back beyond the view to chains of state transitions forming trajectories from the known initial state to possible finite states. The stochastic simulation of trajectories produces objective function values for possible future states. These values can be learned offline by means of simulation and can support online decision making while vehicles are operating on the road.

While already in operation, one may take decisions by relying solely on the offline information provided. Additionally, one may perform short online look-ahead simulations in order to adjust as elaborately as possible to the actual situation faced. Online look-ahead suffers from the relatively small number of simulations to be carried out at the time when a decision is demanded. Thus, offline as well as online approaches have their virtues. Offline approaches are restricted to a coarse grained state representation but depict the global decision space. Online approaches model the actual decision situation in detail but are restricted to a narrow scope of possible future outcomes. Compared to the vast number of publications in the area of static vehicle routing, both approaches have only received minor attention so far.

The book at hand treats subsequent decision making in routing in novel and innovative ways. Marlin Ulmer introduces the vehicle routing problem, stresses the need for a dynamic and stochastic problem formulation and describes sources of uncertainty for vehicle routing. Of particular importance is the in-depth consideration of approximate dynamic programming. Although much detail is presented for the interested reader, Marlin Ulmer follows a clear line of argumentation presenting all this material in terms of the notion of anticipation. Computational experiments elaborate on a routing problem with stochastic customer requests providing evidence for the usefulness of the approach. This work constitutes a milestone in the research of dynamic and stochastic vehicle routing.

Braunschweig, Germany — Dirk Christian Mattfeld
May 2016

Acknowledgements

Within the five years the content of this book evolved, I was accompanied by many friends and colleagues, who contributed to my research and supported me over the time. I want to thank all of them heartily. There are some persons having a special share in my research and life. In the following, I want to thank them individually.

First of all, I would like to thank my adviser Prof. Dirk Mattfeld for his support, his constructive remarks, and the liberty he granted me in my research. My thanks go to the entire Decision Support Group of the Technische Universität Braunschweig and especially to my former colleagues and now friends Prof. Jan Ehmke and Dr. Patrick Vogel for being always available for discussions and giving advice. They all contributed to this work in many ways. I additionally would like to express my gratitude to Prof. Stefan Voß, Prof. Justin Goodson, and Prof. Barrett Thomas for their fruitful comments and the interest they showed in my work.

For me, scientific research is not possible without accompany of family and friends. Therefore, I would like to express my gratefulness to my entire family and especially to my parents Marianne and Ulrich for their wholehearted encouragement and trust throughout my life. I finally would like to thank the most important person, my partner in life Ulrike Roßwag for her patience, understanding, and incredible support in the last years.

Braunschweig
December 2016

Marlin Wolf Ulmer

About this Book

This book is the result of manifold discussions with researches from two different research fields: *vehicle routing* and *stochastic dynamic optimization*, namely, approximate dynamic programming (ADP). Both fields have a long history and can draw a large body of research and methods. While in vehicle routing the demand for stochastic dynamic optimization methods increases, a strong connection to stochastic dynamic optimization (SDO) and ADP is still missing. The purpose of this book is to build a bridge between these fields enabling the broad application of ADP in the field of dynamic vehicle routing.

This book provides a straightforward overview for every researcher interested in stochastic dynamic vehicle routing problems (SDVRPs). The book is written for both the applied researcher looking for suitable solution approaches for particular problems as well as for the theoretical researcher looking for effective and efficient methods of stochastic dynamic optimization and approximate dynamic programming. To this end, the book contains two parts. In the first part, the general methodology required for modeling and approaching SDVRPs is presented. We present adapted and new, general anticipatory methods of ADP tailored to the needs of dynamic vehicle routing. Since stochastic dynamic optimization is often complex and may not always be intuitive on first glance, we carefully accompany the theoretical SDO-methodology with illustrative examples from the field of SDVRPs. The book contains more than 50 explanatory figures. Still, we have been mindful to maintain a succinct red thread through the first and second part of the book. To this end, the theoretical methodology is self-contained and the arrangement of the chapters allows to skip examples, if desired. At the end of each chapter, we briefly summarize the main implications and the resulting steps to expect in the subsequent chapter.

The second part of this book then depicts the application of the theory to a specific SDVRP. The process starts from the real-world application. We describe a SDVRP with stochastic customer requests often addressed in the literature (e.g., Bent and Van Hentenryck 2004; Thomas 2007). We show in detail how this problem can be modeled as a Markov decision process and present several anticipatory solution approaches based on ADP. In an extensive computational study, we show the advantages of the presented approaches compared to conventional heuristics. To allow deep insights in the functionality of ADP, we present a comprehensive analysis of the ADP-approaches.

Further highlights of this book:

- This book gives a comprehensive overview over the real-world applications demanding for anticipatory dynamic decision making. To this end, we present many theoretical and practical sources highlighting the importance of decision support for SDVRPs. We especially identify and motivate same-day delivery, shared mobility, healthcare, and demand responsive passenger transportation as promising future research areas.
- We present a comprehensive literature review and analysis regarding SDVRPs. This review extends current reviews by Pillac et al. (2013); Ritzinger et al. (2015) and Psaraftis et al. (2015) by an analysis of the SDVPRs with respect to problem characteristics and modeling as well as the degree of anticipation provided by the applied solution approaches.
- Besides showing how SDVRPs can be modeled as Markov Decision Processes (MDPs), we give an overview how uncertain events can be modeled as stochastic information. To this end, we analyze the literature and present the general ways uncertain customer requests, demands, travel times, and services times are modeled. We further present examples of MDP-models for these four drivers of uncertainty.
- Since the straightforward application of ADP to SDVRPs is challenging, we adapt general ADP-methods to the requirements of SDVRPs. We further present new ADP-methods, not only suitable for SDVRPs, but for many problems with a complex MDP-structure. In a comprehensive computational evaluation, we not only show the superiority of these methods compared to conventional heuristics, but also present a profound analysis to reveal their advantageous and functionality in detail.

We are positive that this book will function as a foundation in the field of stochastic dynamic optimization for stochastic dynamic vehicle routing problems. It will connect the fields of SDVRP and SDO and will enable researchers to identify and apply suitable solution approaches leading to high quality anticipation in dynamic vehicle routing.

Contents

Acronyms

ADP	Approximate Dynamic Programming
AHS	Ad Hoc Sampling
AI	Anticipatory Insertion
ATB	Anticipatory Time Budgeting
AVI	Approximate Value Iteration
B2C	Business to Customer
BA	Business Analytics
BMD	Brownian Motion with Drift
C2C	Customer to Customer
CBH	Cost Benefit Heuristic
CI	Cheapest Insertion
COG	Center of Gravity
DLT	Dynamic Lookup Table
DTSP	Dynamic Traveling Salesperson Problem
ERC	Early Request Customers
GRASP	Greedy Randomized Adaptive Search Procedure
IZS	Fully Sequential Procedure for Indifference Zone Selection
LM	Logistics Management
LRC	Late Request Customers
LSP	Logistic Service Provider
LT	Lookup Table
MDP	Markov Decision Process
MIP	Mixed Integer Program
MSA	Multiple Scenario Approach
MVF	Approximately Modeling the Value Function
OR	Operations Research
PA	Prescriptive Analytics
PDS	Post-Decision State
PTSP	Probabilistic Traveling Salesman Problem
RA	Rollout Algorithm

RVRP	Rich Vehicle Routing Problem
SDO	Stochastic Dynamic Optimization
SDP	Stochastic Dynamic Program
SDVRP	Stochastic Dynamic Vehicle Routing Problem
SLT	Static Lookup Table
TDTSP	Time-Dependent Traveling Salesperson Problem
TSP	Traveling Salesperson Problem
VFA	Value Function Approximation
VRP	Vehicle Routing Problem
WAE	Wait at End
WAS	Wait at Start
WLT	Weighted Lookup Table

List of Figures

List of Tables

Algorithms

Chapter 1
Introduction

The amount of goods and passengers to transport as well as calls for physical services at customers increase worldwide, especially within urban areas. Transportation and services are conducted by commercial fleets of vehicles. These fleets are managed by logistic service providers (LSPs). LSPs play a major economical role worldwide. In 2014, the overall revenue in logistics reached one trillion dollars in Europe (Fraunhofer 2015). LSPs also are major employers. For example, in truck transportation, around 1.5 million persons are employed in the United States (Burger et al. 2015) today.

The settings for LSPs are recently changing. In particular, expectations of shippers and customers increase. On the one hand, shippers and customers expect reasonably priced services. On the other hand, they expect the services to be fast and reliable. To allow reliable services and to keep the customers loyalty, LSPs need to consider conditions and requirements, e.g., customers' time windows for attended home delivery (Campbell and Savelsbergh 2006), quick responses to the shippers' requests (Hvattum et al. 2006), and overall short delivery times (Lowe et al. 2014). Because LSPs operate in a highly competitive market with low profit margins (Esser and Kurte 2015), they need to balance a tradeoff between costs and reliability. Especially transportation within urban environments is costly and contributes to a high percentage of the entire supply chain costs (Gevaers et al. 2011).

The environment LSPs operate in is highly uncertain, especially serving private customers in urban areas. The limited infrastructure combined with varying street loads may lead to uncertainty in the required *travel times* to commute between different shippers, customers, and the service provider's depot (Kenyon and Morton 2003). The time to deliver and pickup goods or passengers as well as to conduct a specific service at a customer depends on the time for parking, loading, and the required service. These according *service times* are often unknown (Laporte et al. 1992). Further, customers may have differing and uncertain *demands*. LSPs need to maintain the customer's inventory sufficiently, e.g., to avoid stock-outs (Trudeau and Dror 1992). In many cases, the customers are unknown before they *request* for

© Springer International Publishing AG 2017

M.W. Ulmer, *Approximate Dynamic Programming for Dynamic Vehicle Routing*,
Operations Research/Computer Science Interfaces Series 61,
DOI 10.1007/978-3-319-55511-9_1

services (Thomas 2007), passenger transport (Häll et al. 2012), pickup (Ulmer et al. 2017b, or delivery of goods (Voccia et al. 2017). As a result, we identify four major practical drivers of uncertainty: travel times, service times, demands, and requests.

Due to the increasing customer expectations, planning for LSPs becomes more complex. They need to plan regarding multiple objectives, many constraints, and uncertainties. In some cases, they need to determine unalterable a-priori plans. To be efficient, these *robust* plans need to consider future developments, e.g., varying service times (Errico et al. 2016). Still, in an increasing number of cases, LSPs are required to replan, i.e., adapt their planning during the day, e.g., to answer new customer requests. In these cases, a priori plans are not suitable and LSPs require *flexible* plans allowing replanning with respect to newly revealed information. In this book, we focus on the latter class of problems. For these problems, the accessible time for replanning is highly limited because customers expect fast responses and detailed information about their service. To approach the complexity, LSPs draw on decision support tools to determine suitable plans. These plans need to be cost-efficient and effective regarding the objectives. They need to be reliable regarding the constraints and flexible regarding uncertainty.

For decision support, LSPs can access increasing computational resources. To allow the application of decision support methods, the according practical applications are scientifically described as *rich vehicle routing problems* (RVRPs) and approached with methods of operations research (OR). A subclass of RVRPs is the class of *stochastic dynamic vehicle routing problems* (SDVRPs).

SDVPRs are vehicle routing problems with two main characteristics. First, for SDVRPs, planning can be repeatedly adapted at several points in the decision horizon (dynamism). Second, the uncertainty in the exogenous process is substantial (stochasticity) demanding to replan and to integrate predictions of potential future developments.

SDVRPs are "rich" regarding the presence of uncertainty and subsequent planning situations during the day. Since the impact of uncertainty is significant, robust a priori planning for SDVRPs is generally not applicable. Plans need to be flexible to allow a reaction to newly revealed information. Still, for many SDVRPs, mere reactions are not suitable (Powell et al. 2000). Anticipation is essential to avoid ineffective decisions. The main idea of anticipation is to incorporate a predictive model of future uncertainty and decision making in the solution approach (Butz et al. 2003). The predictive model can, e.g., be derived by analyzing historical data. The derivation is provided by the field of *predictive analytics*, e.g., by intelligent data analysis (Ehmke 2012). Classical (deterministic) OR-methods are not able to include predictions but only allow optimization based on the current information. Recently, the field of *prescriptive analytics* has started to close the gap between classical OR-methods and predictive analytics (Evans and Lindner 2012).

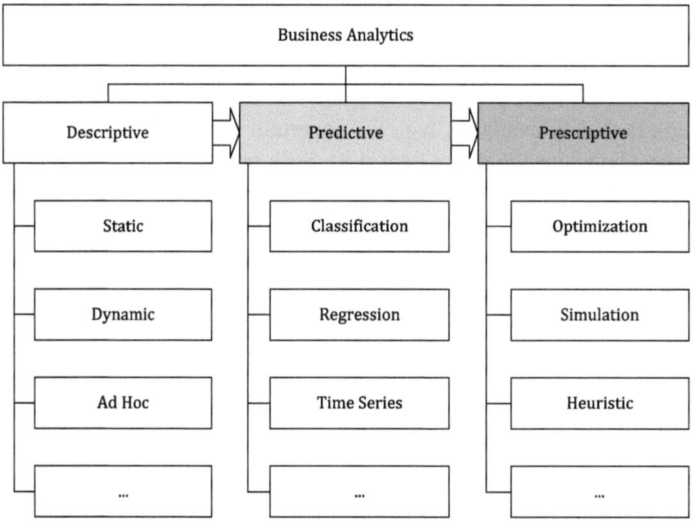

Fig. 1.1 Business analytics

1.1 Prescriptive Analytics

In this section, we describe the concept of prescriptive analytics and embed planning for SDVRPs in this context. Figure 1.1 shows a taxonomy of business analytics (BA) developed by **INFORMS**. In the following, we describe the components of Fig. 1.1 based on (Delen 2014 p.16f). In a nutshell, BA explores an organization's data for decision support. The first basic step of BA is to depict the data based on reports. This is conducted in the field of *descriptive analytics*, often used synonymously with the term "business intelligence". The reports summarize business activities depicting key business indicators like revenues or inventory levels. Reports may be statically, dynamically, or ad hoc meaning they base on individually designed queries. In simple terms, methods of descriptive analytics answer the question "What happens?". The next step of BA is to draw on the findings of descriptive analytics to derive predictions about future developments. This is achieved in the field of *predictive analytics*. Predictive analytics draw on analytical methods like data mining for classification, regression, or time series analysis to answer the question "What will happen?". As an example, expected future customer demands are derived by analyzing the development of previous customer demands. In many cases, predictive analytics provides stochastic information about potential future developments, e.g., mean values and variances of expected future demands. The final step in BA is PA drawing on the predictions to derive anticipatory quantitative decision support, i.e., to determine an action or plan given a specific planning situation. To this end, the stochastic information may be included in an optimization model, used for simulation, or may be used to derive heuristic solution approaches answering the question "What should I do?".

1.2 Scope of This Work

In the field of SDVRPs, predictive analytics is frequently applied, i.e., stochastic information is widely available, e.g., for uncertain travel and service times, or customer request behavior (compare Sect. 4.5). Still, prescriptive analytics is not established for SDVRPs yet. In this book, we focus on PA for SDVRPs. Therefore, we identify practical routing applications containing uncertainty and the requirement for replanning, i.e., flexible plans. We focus on road-based, urban routing applications, mainly for servicing private customers. We show how uncertainty and replanning can be integrated in the scientific model of the SDVRP. We define anticipation and provide a classification of anticipatory solution approaches from the field of PA. We especially consider *reactive anticipation* including the expected reactions of the system to a decision in the decision making process. For reactive anticipation, we give an overview on approaches focusing on methods of approximate dynamic programming (ADP, Powell 2011) and provide two extensions of current ADP-methods to enable efficient and effective anticipation. We use the classification and the presented approaches to provide a thorough literature overview on SDVRPs regarding the four main drivers of uncertainty and the degree of anticipation of the applied solution approaches.

In both practical applications and literature, we identify customer requests as the most prominent driver of uncertainty resulting in significant adaptions of plans. Therefore, we show the process of modeling and how anticipatory solution approaches can be derived for an exemplary SDVRP with stochastic requests. We show how anticipation influences decision making and how reactive anticipation is able to capture the complexity of the SDVRP.

This book aims on establish PA in the field of SDVRPs. Our contributions are manifold. First, we provide a classification for real-world routing applications regarding the degree of uncertainty and requirement for replanning. Second, we give guidance how to scientifically model SDVRPs as Markov decision processes (MDPs, Bellman 1956). We further give an elaborate classification of anticipation and the literature in the field of SDVRPs. Methodologically, we present new generic anticipatory methods in the field on approximate dynamic programming increasing the degree of anticipation and the solution quality of decision making. For the exemplary SDVRP, these methods allow to develop approaches significantly outperforming state-of-the-art heuristics and resulting in high quality practical decision making.

1.3 Outline of the Following Chapters

The outline of the book is depicted in Fig. 1.2. This book contains two parts. The first part gives a general overview on SDVRPs, describing the process starting from the real-world application requiring replanning and consideration of uncertainty, over scientific modeling, to anticipation and anticipatory solution approaches.

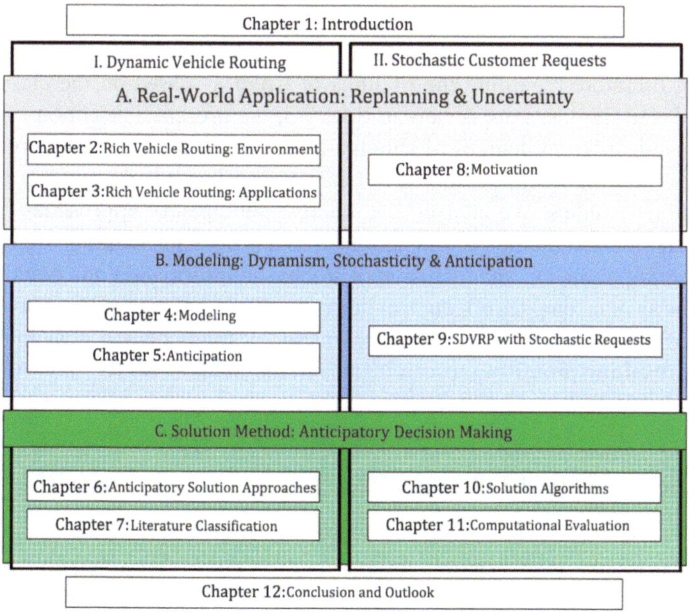

Fig. 1.2 Outline of this book

The second part shows the implementation of this process for an SDVRP with stochastic customer requests. Every part is divided into three segments. We start with the practical application. We then depict how the practical application can be modeled scientifically as SDVRP or MDP respectively. We finally show how the SDVRPs can be approached by anticipatory solution methods.

In the first part, we define and classify practical routing applications, in the literature known as RVRPs. We focus on a subset of RVRPs, SDVRPs and on road-based routing in an urban environment. We first analyze the (business) environment inducing RVRPs in Chap. 2 and justify the assumptions about the main drivers of uncertainty and requirement for replanning. We further describe the technical advances enabling replanning and anticipation. The application fields for SDVRPs are classified in Chap. 3 and the specific applications are distinguished explicitly requiring replanning and anticipation. We model uncertainty as stochasticity, replanning as dynamism and the SDVRPs with respect to an MDP in Chap. 4. To include stochastic information in decision making, an elaborate classification of anticipation is presented in Chap. 5. We provide an overview of general anticipatory approaches for different degrees of anticipation in Chap. 6. For the different drivers of uncertainty, we give an extensive literature classification regarding work on SDVRPs in Chap. 7 focusing on the modeling and on the degree of anticipation of the applied approaches.

In the second part, we embed an exemplary SDVRP with stochastic customer requests in the presented classifications. After a motivation of the problem and of the requirement for anticipation and replanning in Chap. 8, we identify the inclusion of spatial and temporal information about stochastic requests as key requirement for anticipatory planning. We model the SDVRP as an MDP in Chap. 9 and analyze the according literature regarding the findings of Chap. 8. Based on the classification in Chap. 5 and the literature review in Sect. 9.3, we present a selection of solution algorithms of different degrees of anticipation in Chap. 10. Therefore, we draw on the general ADP-methods presented in Chap. 6. In Chap. 11, we conduct extensive computational studies. We analyze the reactive anticipatory approaches regarding their effectiveness and efficiency in anticipation and decision making. We show how reactive anticipation, i.e., the inclusion of stochastic information and potential reactions of the system, enables effective anticipatory decision making. In the conclusion and outlook, we summarize the main statements of the book and give an overview about the resulting promising research areas in the fields of RVRPs and ADP.

1.4 A Recipe for ADP in SDVRPs

In the following, we present a brief recipe how to address SDVRPs with ADP. If you are thinking about approaching a potential SDVRP-application, this recipe may assist you in your journey through the three steps of problem identification, modeling, and definition and application of an anticipatory ADP-method. There is no unique way to approach an SDVRP, but our experience with our own research, with graduate students, and PhD candidates shows that taking the following steps may be helpful to get started and to eventually achieve anticipatory policies. In each of the following steps, we refer to the according chapters of this book.

1.4.1 The Application

First of all, you have to determine whether the underlying application induces an SDVRP. Even though nearly every VRP-application can be seen as an SDVRP, in some cases, the consideration of stochasticity and dynamism may not be necessary. In Chap. 2 the environment is described, SDVPRs often occur in. In Chap. 3, we give a comprehensive analysis of potential application fields, e.g., same-day delivery. Still, the applications inducing SDVRPs are manifold. The questions to ask are the following: Is the problem stochastic, i.e., do uncertainties significantly impact the applications outcome? If the uncertainties are neglectable, you may decide to consider the application as a deterministic VRP. A deterministic VRP is by definition static. If the uncertainty is significant, you should identify the main driver of uncertainty. This may be the presence of uncertain customer requests, the customers' demands, et cetera. For applications with manifold stochastic impacts, you should again ask for the significance of their impact. In some cases, uncertainty in travel times may significantly impact the rewards or costs. In other cases, this uncertainty may not impact the overall decisions' outcomes. If you have decided that the application induces a stochastic VRP, you have to determine whether the VRP is dynamic, i.e., whether the possibility (and the requirement) to react to newly revealed, previously uncertain, information is given. For some applications, reactions are not possible, e.g., in cases, where the dispatcher commits to the initial decision or where dispatcher and drivers do not communicate once the drivers started their shifts. These problems are stochastic but static. For other applications, dynamic decision making is mandatory, e.g., when new customers are assigned to vehicles during the day. If you decide that dynamic decision making is relevant for your application, you have identified an SDVRP.

To compare your SDVRP to the literature, you find a comprehensive analysis of more than 50 publications on SDVRP in Chap. 7. This analysis lists the SDVRP-literature with respect to the drivers of uncertainty as well as the objective of the application. Generally, the objective is to exploit limited resources to maximize rewards or to fulfill a set of task minimizing costs. The literature may further assist you in the modeling and the determination of a suitable solution approach. To this end, we give an overview on the models and the applied solution approaches for the publications.

1.4.2 The Model

If you decide that the application induces an SDVRP, the problem has to be modeled as an MDP. Notably, the model is independent of any approach you may apply to the problem. In Chap. 4, we give an overview on the required components of the MDP-model as well as exemplary models for different basic SDVRPs. Further, in Chap. 7, we present how potentially similar problems are modeled in the literature. You may be able to modify an existing MDP from the literature. To model your particular SDVRP, you have to model decision points, states, decisions, rewards or costs, and transitions. First of all, you have to determine, when decisions are made, e.g., in real-time, every minute, or only when an event occurs, e.g., when a new customer requests service. Second, you have to define the according decision state. This state is unambiguously and contains all the necessary information available at a decision point, e.g., point of time, vehicle locations, but also possible information about future developments, i.e., forecasts. If the state is defined, the set of possible decisions can be determined. Notably, a decisions only determines the "next step" of the problem. To this end, you should identify the set of (unalterable) measures possible in a decision point. As an example, a decisions is made often not about the entire sequence of customers, but only about the next customer to visit. Further, you have to determine the resulting rewards and costs of a decision. In some cases, they only reveal after the stochastic part of the transition and are therefore random variables and difficult to capture.[1] Finally, you have to ensure feasibility with respect to the constraints like time windows or capacities.

Our experience in the field of SDVRPs showed that splitting the overall transition between two states in a deterministic decision part and a stochastic transition part significantly simplifies the modeling. As a result, it may be helpful to model the post-decision state explicitly. This post-decision state is often similar to the decision state. If the problem, e.g., demands for assignment decisions of new customer requests, the decision state contains the open requests while the post-decision state only contains the resulting assigned customer and no requests. The stochastic part of the transition significantly depends on the problem under consideration. It may be, e.g., the realization of an uncertain customer demand, the finish of a service at a customer with before uncertain service time, or the occurrence of new requests. We present an analysis and instructions with respect to the stochastic transition for prominent SDVRP-problem classes in Chap. 4. Finally, you have to determine the initial and the termination state of the problem, i.e., when and how does the problem start and end. Usually, in the initial state, the vehicles are located at the origin, e.g., the depot and no customer is assigned yet. In the termination state either all customers are served or the time limit is reached depending on the SDVRP. In the end, you have an unambiguous mathematical model of your SDVRP.

1.4.3 Anticipatory Approaches

There is no general solution approach for every possible SDVRP. Still, there are general ways to identify suitable approaches. If you already have an approach in mind, you may skip this section. Still, in the following, we give guidance based on our experience, how we approach SDVRPs.

First of all, a simulation framework is required to analyze potential solution approaches. Fortunately, the MDP-model already provides a straightforward structure of this framework with respect to decision points, states, decisions, and stochastic transitions. Compared to static VRPs, understanding and capturing the dynamic behavior of SDVPRs is rather complex. We made the experience, that the implementation of an initial simple policy is really helpful to get a feeling for the behavior of the SDVPR. This policy may be a rule of thumb or even be myopic, i.e., maximizing the immediate reward in every decision point or minimize the immediate costs respectively. This policy neglects all potential future rewards or costs. As a result, this myopic policy either consumes a high amount of the resources for rewards in the first decision points and/or postpones a high amount of costs to later decision points. This myopic policy may function as a first benchmark policy for your SDVRP.

The next step is to find a policy avoiding myopic behavior by means of anticipation. Generally speaking, an anticipatory policy maintains flexibility in the resources and

[1] We recently introduced a route-based MDP model particularly suitable for SDVRPs (Ulmer et al. 2016a). We motivate and summarize these models in Chap. 4.

avoids the decisions leading to expected high costs later. We give a comprehensive definition of anticipation in Chap. 5. In our research, we have identified and analyzed manifold methods to achieve anticipatory policies. We have experienced that the complexity of SDVRPs demands for methods able to adapt to the SDVRP's structure. These methods generally base on approximate dynamic programming. The idea behind ADP is to evaluate a post-decision state with respect to an estimation of the expected rewards, the so called *values*. If you have access to the values, you can select the decision maximizing the sum immediate reward and resulting value. The estimation of the values is usually achieved by means of simulation. In the following, we describe three general methods of ADP suitable for SDVRPs. We define these methods in Chap. 6 in detail.

The first, probably most intuitive ADP-method for an SDVRP is the post-decision rollout algorithm (RA). We give a comprehensive definition and examples for the RA in Sect. 6.3. Given a state, a set of possible decisions, and the set of resulting post-decision states, for every post-decision state, the RA simulates the future decision points. The value is then estimated by the (average) realized rewards within the simulations. Since decision making within the simulations is required, the RA draws on a *base policy*. In our case, this could, e.g., be the myopic policy. The RA can be easily applied to nearly every SDVRP but has one major shortcoming. Since the RA is applied online in real-time, the calculation time is limited and only a small number of simulations is possible. In Sect. 6.3, we give directions how the limited time can be used effectively. Still, if the runtime for simulations is high, you may think about cutting the simulations, e.g., after a number of decision points.

There is a way to avoid the extensive online simulations and to shift the majority of calculation to *offline* simulations. This can be achieved by identifying and exploiting dependencies between post-decision state attributes and values and hence, to derive a function mapping post-decision states to values. This function bases on post-decision state attributes. The specific dependencies are then approximated by means of offline simulation. Once the approximated functions are derived, they can efficiently be accessed online. The method is known as value function approximation (VFA). There are two ways of VFA: actually approximation of the entire function (e.g., via regression methods) or approximation of post-decision states, called approximate value iteration (AVI). Our experiments and multiple discussions with other researchers revealed that for SDVRPs the approximation of the entire function often does not provide good results. This can be explained by the complex dependencies between post-decision state attributes and values, i.e., the complex value function structure. The approximation of the value function with some "regular" functions is often insufficient. We experienced that AVI is more suitable to capture the complex structure, since, instead of the function, single values are approximated individually. Hence, AVI allows an unbiased approximation. We describe AVI in detail in Sect. 6.5.

Since the number of post-decision states is usually vast, a storage of an individual value for every single post-decision state is not possible. To apply AVI, a state space aggregation is generally necessary. This aggregation represents a post-decision state by a set of attributes. Finding suitable attributes is the major challenge for AVI. On the one hand side, the aggregation requires low dimensionality to allow the storage of the values. Hence, the number of attributes is limited. On the other hand, the aggregation should maintain the post-decision state's features relevant for a proper value estimation. There are several ways to determine suitable attributes. You may think about attributes generally differentiating the states and impacting the expected future rewards, e.g., the point of time. The later it is in time, the less rewards can be expected. To evaluate, whether an attribute is suitable, it may be helpful to analyze the impact of an attribute to the value with respect to the RA or the myopic policy. We have experienced that numerical or ordinal, temporal attributes are often well suited while spatial attributes are often not able to reflect the complex dependencies between states and values. Given stochastic requests, an efficient representation of the available resources in a state can be, for example, the amount of slack per vehicle, i.e., the amount of remaining free time budget. For stochastic demands, a suitable representation may the vehicle's inventory level. Further, attributes should represent states leading to high expected costs in the future. If you are aiming for minimizing expected travel times, this may be the number of customers still to serve, the average or maximal expected travel duration to each of these customers, or the average expected travel distances between these customers, et cetera. As you see, the options are manifold and the selection for a specific SDVRP not always self-evident. For an SDVRP with stochastic requests, we show the process how to derived suitable attributes in Chap. 10.

If you have identified a set of promising ordinal or numerical attributes, you are able to aggregate the original state space to a low-dimensional vector space. If the attributes are discrete, you may apply AVI directly on the vector space. Still, often the number of vectors is high leading to inefficient and distorted approximation. It may be effective to partition the vector space to a lookup-table (LT), e.g., by statically partition each numerical dimension in equidistant intervals. The values are then approximated for each entry. In many cases, there is significant potential of improvement for the static partitioning. To this end, we propose the dynamic lookup-table (DLT) in Sect. 6.5. The DLT adapts the partitioning to the approximation process and has already proven to provide superior approximation with respect to quality, runtime, memory consumption, and scalability. Recent experiments show that the DLT is able to efficiently handle discrete vector spaces with four dimensions and more than 50 million different vectors without limitations.[2] You may experience that your attribute selection is not suitable to represent states. In this case, it may be useful to change, add, or remove attributes. If you were eventually able to identify suitable attributes for AVI or even for VFA, you are lucky for two reasons. First, you have already found an efficient, high quality policy for your problem. Second, you can additionally integrate the policy in a *hybrid* RA as a base-policy. We define the hybrid RA in Sect. 6.6. We have experienced (and proven) that for many stochastic dynamic decision problems, the hybrid RA improves both the online and offline

[2] Please feel free to contact us if you are interested in an efficient implementation of the DLT and an overview on the problems the DLT is successfully applied to.

ADP-method. We show the reasons in detail for our SDVRP in the computational evaluation in Chap. 11.

One final step: If you have followed our recipe, please let us know. We are highly interested in your ideas and approaches with respect to the aggregation and the partitioning. We are particularly interested in the obstacles on your way and in your suggestions for improvement. Thank you and good luck!

Part I
Dynamic Vehicle Routing

Chapter 2
Rich Vehicle Routing: Environment

In this chapter, we first recall the development in the field of vehicle routing problems (VRPs). We then describe the characteristics of rich vehicle routing problems (RVRPs) in Sect. 2.2 focusing on uncertainty and replanning aspects. We briefly describe logistics management (LM) in Sect. 2.3 and embed (rich) vehicle routing in context of LM's hierarchical planning in Sect. 2.4. We describe the environment rich vehicle routing is conducted in, the recent developments, and emerging challenges in Sect. 2.5. In Sects. 2.5.1–2.5.3, we identify travel times, service times, customer demands, and requests as the four main drivers of uncertainty. We further identify time windows, working hours, and capacities as the major constraints and costs and reliability as the major objectives for RVRPs induced by the practical applications.

Uncertainty results in the requirement for replanning. Further, a consideration of possible future events in current planning is desired. Therefore, we describe the technologies enabling replanning and stepwise planning as well as predictions of uncertain future events in Sects. 2.5.4 and 2.5.5. Since the majority of routing applications is road-based and uncertainty and the requirement for replanning occur especially in conurbations, we focus on road-based routing in urban areas.

2.1 Vehicle Routing

Vehicle routing reflects the quantitative scientific field for servicing customers by a fleet of vehicles. Vehicle routing problems have a long tradition. As first VRP the classical traveling salesperson problem (TSP) can be seen. In the TSP, a salesperson visits a set of customers. The objective is to determine a *tour* visiting every customer with minimal overall travel costs. As the TSP, the vast majority of VRPs are considered static and deterministic meaning that all information is known a-priori. These VRPs are usually modeled via mixed integer programs (MIPs) based on a mathematical graph. These graphs consist of a set of vertexes and a set of connecting edges. The vertexes represent customers or depots. The edges represent the paths between

© Springer International Publishing AG 2017

M.W. Ulmer, *Approximate Dynamic Programming for Dynamic Vehicle Routing*,
Operations Research/Computer Science Interfaces Series 61,
DOI 10.1007/978-3-319-55511-9_2

customers and depots. Edges usually contain costs. These costs depict the travel times or distances between customers. For VRPs, a solution generally is defined by the assignment of customers to vehicles and the sequence of the assigned customers for every vehicle, i.e., the classical idea of "routing".

Solutions for static and deterministic VRPs are derived with methods of combinatorial optimization. Optimal solutions are achieved by means of branch and bound (Land and Doig 1960), branch and cut algorithms (Nemhauser and Wolsey 1988), or dynamic programming (Christofides et al. 1981). The derivation of solutions for VRPs is generally of non-polynomial complexity (Garey and Johnson 1979), i.e., a small increase in the instance's size results in a significant increase in the solution space. Optimal solutions can only be obtained for instances of small size. For larger instances, solutions are derived by a "good guess" of how an optimal solution may look like. These approaches are called *heuristics*. A classical heuristic for the sequencing of customers per vehicle is, e.g., cheapest insertion by Rosenkrantz et al. (1974). In the last decades, more elaborate heuristics have emerged. These *metaheuristics* systematically explore the VRP's solution space for a near-optimal solution (Blum and Roli 2003). Metaheuristics often combine (optimal) MIP-solution techniques with heuristics leading to so called matheuristics (Hansen et al. 2009). Matheuristics generally divide the global problem in smaller sub-problems. The sub-problems are selected heuristically and often solved to (local) optimality. In essence, the body of research and methods for static and deterministic VRPs is vast. We refer the interested reader to Toth and Vigo (2001) for a detailed overview on VRPs.

2.2 RVPR: Characteristics and Definition

Compared to VRPs, work on rich vehicle routing focuses more on real-world routing applications. These practical applications are "rich" with respect to the objectives, constraints, and uncertainties induced by the real-world routing environment (Caceres-Cruz et al. 2014). As a result, quantitative decision support for rich vehicle routing problems is challenging compared to theoretically motivated vehicle routing problems. A major impediment for the application of quantitative decision support methods to RVRPs is the uncertainty, vehicle dispatchers have to face in their planning. In our analysis of rich vehicle routing, we focus on the uncertainty and the resulting requirement for replanning based on uncertain events as well as the integration of uncertainty in planning.

A single, unambiguous definition of RVRPs is not yet available. Lahyani et al. (2012) describe RVRPs to be "either a VRP that incorporates many strategic and tactical aspects and/or a VRP that reflects the complexities of the real-life context by various challenges revealed daily." RVRPs usually contain complexity in one or several of the following aspects: customers, vehicles, infrastructure, objectives, and constraints. For example, customers may have time windows, vehicles may have loading capacities, or the objectives may contain multiple criteria. For an extensive taxonomy of RVRPs, the interested reader is referred to Lahyani et al. (2015).

In the following, we focus on two specific aspects of RVRP, uncertainty and the requirement for replanning. Caceres-Cruz et al. (2014) state that "dynamic VRPs (so-called real-time VRP) can be also considered as part of the overall RVRP scope." In this context, *dynamic* means that subsequent decision points are considered in contrast to *static* problems with only a single (a priori) decision . Subsequent decision points are especially common if "uncertainty over some variables" is given forcing dispatchers to adapt their current plans. Lahyani et al. (2015) define these types of RVRPs as dynamic and *stochastic*: "The *deterministic* routing problem assumes that the problem parameters are known with certainty while the *stochastic* data assumes that probability distributions are associated with them." In essence, dynamism and stochasticity are two aspects distinguishing many RVPRs and impeding the application and the performance of conventional quantitative approaches designed for plain VRPs (Powell et al. 2000).

2.3 RVRPs in Logistics Management

In this section, we embed rich vehicle routing as part of company's logistics management. In 1986, the Council of Logistics Management defines logistics management as follows (Stock and Lambert 2001).

"Logistics management is an integrating function which coordinates and optimizes all logistics activities, as well as integrates logistics activities with other functions. ... Logistics management ... plans, implements, and controls the efficient, effective forward and reverse flow and storage of goods, services, and related information between the point of origin and the point of consumption in order to meet customers' requirements."

In context of LM, vehicle routing is conducted to deliver or pick up goods and to conduct services at customers. LM installs and controls the infrastructure required for vehicle routing. This concerns the construction and maintenance of storage facilities like warehouses or depots and the according inventory management. Further, LM provides, manages, and maintains the fleet to conduct routing. LM is also responsible for the planning of routes and assignment of customers to vehicles. Stock and Lambert (2001) extend the definition by explicitly adding *forecasting* and therefore the anticipation of future events like customer demands to the tasks of LM. We define forecasting and present the available tools in Sect. 2.5.5. First, we classify vehicle routing in LM's decision making hierarchies and give an overview of the recent developments in real-world routing applications in the context of LM and vehicle routing.

Table 2.1 Hierarchies

Level	Impact	LM	Decisions
Strategical	Long term	Network	Infrastructure
Tactical	Medium term	Service network	Constraints
Operational	Short term	Vehicle routing	Assignment, sequence

2.4 RVRPs in Hierarchical Decision Making

Hierarchical decision making differentiates three levels: strategical, tactical, and operational (Schneeweiss 1999, p. 109ff). Crainic (2000) gives an overview of the management levels regarding LM and transportation. Table 2.1 shows the three levels and their impact on the logistical processes.

On the strategical level, long term decisions are made considering months and years in planning. The planning is conducted on highly aggregated information. Strategical decisions are expensive and not easily alterable. On the tactical level, medium term planning is conducted for weeks and months. Decision making is still performed on aggregated information. On the operational level, strategical and tactical plans are implemented in short term. The operational level consists of daily planning on highly detailed information.

In the context of logistics, the three levels can be differentiated as follows. On the strategical level, decisions are made especially regarding the *infrastructure*, i.e., the depot(s) and the fleet of vehicles. This is often conducted by network planning (Salvendy 2001, p. 1472ff). On the tactical level, *constraints* as driver shifts, vehicles' maintenance strategies, and the specifications of the products (e.g., goods or services) are defined. Transportation flows in the service area are determined. Product specifications may be, e.g., prices for services or deliveries, but also the characteristics of time windows or working hours. Planning on the tactical level is conducted by service network design.

Crainic (2000) defines the tasks on the operational level in transportation applications as "the implementation and adjustment of schedules for services, crews, and maintenance activities, the routing and dispatching of vehicles and crews; the allocation of scarce resources." Dispatchers plan "in a highly dynamic environment where the time factor plays an important role and detailed representations of vehicles, facilities and activities are essential." On the operational level, the vehicles are dispatched to serve customers given the infrastructure, fleets, and constraints like driver shifts. Dispatchers implement routing plans considering infrastructure, resources, time, and constraints. Routing consists of assignment and sequencing decisions and is mainly planned on the operational level. Customers are assigned to vehicles and the sequence of the assigned customers is determined for every vehicle. Routing is often conducted under uncertainty. In many cases, replanning and adaptions of the plans are required.

Work on RVRPs focuses on the operational level of LM. Nevertheless, dependencies exist between the three planning levels. The given infrastructure and constraints influence the resulting routing decisions. In many cases, the efficiency of strategical and tactical decisions are strongly connected to the applied routing. As a result, many decisions on infrastructure and constraints consider routing. As an example, operational assignments are often induced by a predefined tactical partitioning of the service area.

2.5 Recent Developments of the RVRP-Environment

In this section, we present the recent real-world developments in the field of LM focusing on the developments' impacts to RVRPs. Driving factors of the increasing demand for RVRP-decision support are e-commerce, globalization, urbanization, demography, and new business models enabled by emerging technologies. For the different factors, we examine the increase in uncertainty and the impact on required replanning. Further, we present technological advances allowing for replanning and forecasting.

2.5.1 E-Commerce and Globalization

E-commerce is the umbrella term for the trading of goods and services via electronic devices, usually over the internet. For 2015, prognoses assume that the worldwide e-commerce transactions will nearly double compared to 2011 (Capgemini 2012). The number of transactions is expected to reach 38.5 billion. The growth of e-commerce leads to an increase in the quantity of transported goods and a change of customers' behavior. In conjunction with e-commerce, globalization is another driving factor for the increasing demand for transportation. Globalization allows for trading between different countries and continents and further increases the number of online orders and the according shipping. In Germany, the combination of globalization and e-commerce results in around 2.8 billion shipped parcels in 2014 (Esser and Kurte 2015).

Besides the increase in required transportation, e-commerce especially changes customers' behavior and expectation. Ordering and selling goods online is simple and convenient. Online shops are always accessible without any closing hours. Selling online does not require an expensive brick-and-mortar store, but only the possibility of fast shipping. In recent years, a vast range of online shops has emerged. Ordering online saves costs, time, and effort for traveling to the stores and waiting in lines (Bubner et al. 2014). After placing the order, customers can track, change, and update the orders via mobile phone at any point of time (DHL 2013). These possibilities lead to more spontaneous, i.e., uncertain customers' consume behavior. Service providers may have to manage their inventory accordingly.

Along with the change of customers' behavior, the expectations rise. In contrast to shopping at stores, customers do not receive the desired good right after placing the order. Instead, they have to wait until the delivery arrives. Studies show that delivery time and delivery costs are two of the main success factors in e-commerce (Lowe et al. 2014). Customers expect reasonable priced and fast delivery. To match these expectations, same-day delivery is already provided in some cities (Wahba 2015) especially in the field of grocery delivery (Campbell and Savelsbergh 2006). For e-commerce companies, same-day delivery is seen as a main factor for future success (Mangalindan 2013). Providing same-day delivery requires constant adaptions of the delivery plans because new uncertain order *requests* occur during the day when the vehicles are already conducting deliveries. A priori planning is not suitable. In many cases, dispatchers are only able to plan stepwise.

2.5.2 Urbanization and Demography

Urbanization describes the process of people moving from rural to urban areas. The process of modern urbanization started in the 1950s. Currently, more than half of the world population lives in a city. The United Nations expect this number to increase up to two-thirds in 2050 (United Nations 2015). Urbanization impacts both the requirement for transportation and physical services at customers as well was the environment the routing has to be conducted in.

Transportation

A result of the urbanization is an increasing demand for goods, passenger transports, and services. The main business area for delivery and service companies is already located in the cities (Jaana et al. 2013). As described in Sect. 2.5.1, the amount of ordered goods increases. Combined with urbanization, this leads to a substantial growth in good transportation within the city. An extensive overview over the according field of *City Logistics* is given by Taniguchi et al. (2001). In 2007, transportation already caused more than 10% of the overall city traffic (Figliozzi 2007). Last-mile delivery is one of the most expensive parts of the entire supply chain (Gevaers et al. 2011). Therefore, effective and efficient routing is essential for service providers' success.

Beside the transportation of goods, the demand for passenger transport increases. The mobility demand is expected to more than double until 2050 (Schafer and Victor 2000). This development will increase the challenges in public transportation, intermodal transportation, and shared mobility. As a combination of public and individual transportation, the use of demand responsive passenger transport (e.g., dial a ride) increases (American Public Transportation Association 2013). This leads to higher uncertainty in planning and requires immediate responses to new customer requests. Dispatchers further have to dispatch vehicles considering the vehicle's *capacities*. Shared mobility systems have become an essential part of urban transportation. More than 800 cities provide bike sharing systems worldwide. The number increases rapidly (Steinsiek 2015). The number of users in car sharing systems has doubled in the last two years (European Automobile Manufacturers' Association 2015). Providers of shared mobility systems have to ensure sufficient vehicles and parking spaces given uncertain customer requests and *demands* (Brinkmann et al. 2015).

Services

Due to the growth in urban population, evidently the number of service calls increases. This is especially the case in the field of healthcare. Due to the demographic developments, the percentage of the population aged 65 and over will double until 2050. In Europe, this age class is expected to represent more than one-fourth of the overall population in 2050 (United Nations 2010). In the United States, the required spending for outpatient and hospital treatments is expected to grow by 40% in the next five years (Economist Intelligence Unit 2013a). Services have to adapt to this aging society, e.g., to meet the demands for (ad hoc) house calls and patient transfers (Bubner et al. 2014). The requests for house calls and the *service time* to spend at a patient are often unknown. Dispatchers have to schedule the physicians accordingly, change schedules regarding the required time at a patient and the physicians' *working hours*, and assign new requests to physicians during the day.

2.5.3 Urban Environment and Municipal Regulations

All these tasks have to be conducted in the limited city infrastructure and with respect to environmental conditions. The capacities of streets in a city are determined by the street size and the traffic control strategy (e.g., traffic lights) applied by the city's administrator, i.e., the traffic management (Taniguchi et al. 2001, p. 4). Street sizes in the cities are mainly unalterable. Generally, the control of the traffic management is only able to increase the capacity of one street by reducing the capacity of another. The limited capacity of the streets combined with an increase in traffic volume leads to congestion within the city. Congestion results in an increase in required *travel time* by up to 50% in Europe (TomTom 2015a) and North America (TomTom 2015b). Beside congestion, drivers have to deal with limited parking and delivery zones (Dablanc 2007). This leads to uncertainty in the amount of time to serve a customer. Dispatchers have to consider the uncertainty in travel time and service time while scheduling tours. Further, they have to adapt the plans according to new traffic information or to changes by the traffic management (Köster et al. 2015).

The increase in traffic has led to substantial emissions, e.g., air pollution, carbon dioxide emissions, and noise. Right after the energy sector, the transportation sector is currently the second largest generator of carbon dioxide emissions in the United States (U.S. Energy Information Administration 2015). Transportation is responsible for a major part of emissions in the city (Organization for Economic Cooperation and Development 2013). The emissions lead to decreases in quality of life and citizens' convenience. As countermeasure, municipalities issue fines or restrict access for vehicles at certain times (Quak and Koster 2009) and due to emission alerts via traffic management (Irvine 2013). The emission alerts are not known, but are influenced by traffic and weather conditions. The traffic management's reactions to these alerts changes the required travel times in these areas. As a result, dispatchers may have to adapt their plans.

2.5.4 *Technology*

As described in Sect. 2.5.1, technologies like smartphones change customers' behavior and enable new business models forcing dispatchers to replan or even plan stepwise. The technological advances can also be used by dispatchers and vehicle drivers for communication and for replanning their routing and schedules. Digitization and the omnipresence of smartphones enables new modes of transportation, e.g., crowdsourced delivery (Barr and Wohl 2013), crowdsourced passenger transport (Huet and Chen 2015), drone delivery (Sinha 2013), or delivery to car trunks (Cartledge 2012; Behrmann and Weiss 2014). These new transportation modes may allow dispatchers more flexibility in planning, but also contain uncertainty in both the customer's location and, for crowdsourced transportation, even the accessible fleet.

At every point of time, dispatchers have access to a real time information process, e.g., vehicles' positions via GPS, customers' inventories (Verma and Campbell 2014), or traffic statuses via geographic information systems and digital road maps. A synchronization between the information and planning process, i.e., the recalculation and replanning to updated information is enabled by an increase and an outsourcing of computational resources, e.g., real-time services (Bubner et al. 2014). As a result, dispatchers are able to immediately react to updated information.

The increase in autonomous logistics allows a fast and accurate provision of goods in the warehouses. Together with combined planning of online retailers and delivery companies, e.g., by cloud based supply chain management, this allows a faster response to customer orders eventually enabling same-day delivery (Leukel et al. 2011). This fast provision of goods and response leads to a reduction of planning time and another increase in uncertainty.

2.5.5 *Data and Forecasting*

A mere reaction to updated information may lead to inefficient planning. It is desirable to plan ahead including forecasts about future events into current plans. Forecasts are derived from experience and knowledge in the field of predictive analytics. This knowledge is mainly hidden in historical data accumulated over time. Recently, decision making based on data gains in importance (Economist Intelligence Unit 2013b). Therefore, data has to be analyzed and included in anticipatory algorithms.

Data Analysis

Computer memory has become cheap. The digital control of devices results in vast amounts of unstructured (big) data (Manyika et al. 2011). With respect to vehicle routing, data is collected regarding the vehicles, the customers, and the goods or services. The collected data implicitly reflects the environmental impacts like weather or traffic statuses. Dispatchers additionally have access to external information and data, e.g., current expected travel times. Amongst others, companies track vehicles' routes, travel times, service times, load, and speed. They further track customers' locations, request times, service times, and the ordered good or service. The collected data may allow forecasts about future events like customer requests or changes in travel time. Nevertheless, the data is often still "untapped" (Bubner et al. 2014).

In order to utilize the data for anticipation of the future, two challenges arise. First, the data has to be transferred to *information*. For this purpose, the large amount of data has to be preprocessed and aggregated. This is conducted in the field of predictive analytics, e.g., by data mining and intelligent data analysis . The interested reader is referred to Hand et al. (2001) and Berthold and Hand (2003) respectively. Second, to use the information in planning and decision making and to allow anticipation, suitable algorithms have to be implemented by the field of prescriptive analytics. In this book, we focus on the second challenge, the derivation and implementation of *anticipatory vehicle routing approaches*.

Anticipation

Anticipation means the inclusion of possible future events like congestion, breakdown of vehicles, or new customer requests into current planning. As an example for anticipation, the police of Los Angeles anticipate potential crimes based on historical data. They schedule their vehicles depending on the daytime and the area (Kelly 2012). Especially, the field of "anticipatory logistics" is emerging (Bubner et al. 2014). Recently, a patent was issued, to anticipate customers' orders before they are actually placed (Spiegel et al. 2012; Kopalle 2014).

Anticipation is achieved in the field of *prescriptive analytics*. Anticipatory planning differs from *myopic* planning. Myopic planning does not consider the future. An anticipatory plan may require more resources and higher costs or may provide less immediate gain, but it is robust or allows adaptions and provides flexibility (Powell et al. 2000). As an example, a robust plan for delivering grocery anticipating uncertain travel times may contain temporal safety buffers (Ehmke et al. 2015). A vehicle may not be able to serve as many customers as following a myopic plan, but it is more likely to meet the customers' time windows. As an example for flexible planning, dispatchers may save an amount of free time budget to serve potential future requests (Ulmer et al. 2017b).

2.6 Implications

As presented in this chapter, the requirement for anticipatory RVRP-solutions rises especially in an urban environment. Dispatchers have to plan with respect to reliability and cost efficiency. The conditions in which vehicle routing is conducted change more and more frequently. Travel times, service times, customer demands, and customer requests are uncertain. These (uncertain) changes force dispatchers to replan with respect to updated information. New technologies allow the immediate reaction to updated information. Nevertheless, the anticipation of the future may be necessary to achieve efficient and effective current plans. An anticipatory plan is not necessarily the immediate "best" plan based on current information, but allows flexibility and reliability regarding the future.

Chapter 3
Rich Vehicle Routing: Applications

In Chap. 2, we have described the environment, vehicle routing is conducted in. In the following, we present the practical fields of routing applications inducing RVRPs. We analyze the applications regarding uncertainty and requirement for planning. We focus on routing in urban environments. The main purpose of this section is to give an overview of the important entities and underlying components in RVRPs as well as the most common objectives, constraints, and main drivers of uncertainty based on the findings of Sect. 2.5. Notably, we focus on the drivers of uncertainty and confine ourselves to the most conventional objectives and constraints. We use the induced classification to present the most prominent practical routing applications. We analyze the different applications regarding the requirement for replanning or stepwise planning.

First, we formally present the required entities, objectives, constraints, and uncertainty factors to describe the majority of vehicle routing applications as depicted in Fig. 3.1. Entities are vehicles and customers depicted in black as well as the surrounding infrastructure depicted in gray respectively. The entities and the underlying components are described in Sect. 3.1. Dispatchers' *plans* contain the assignments of customers to vehicles and the sequencing of the assigned customers per vehicle. This is indicated by the double arrow. Plans are described in Sect. 3.2. Objectives are depicted in white and are twofold. Providers aim on reliable customer service and cost efficient routing as described in Sect. 3.3. The constraints are customers' time windows, drivers' working hours, and vehicles' capacities. The constraints are depicted in the top of Fig. 3.1 and are described in Sect. 3.4. The main drivers of uncertainty are vehicles' travel times, customers' requests, demands, and service times. They are depicted in bottom of Fig. 3.1 and are described in Sect. 3.5. Temporal constraints and uncertainty factors are striped vertically. Capacity constraints and the according potentially uncertain demand are depicted with horizontal stripes.

The entities combined with the Cartesian product of objectives, constraints, and uncertainty factors span a field of possible routing application allowing a classification of practical routing applications. We use this classification to present an overview of the most prominent practical applications requiring routing in Sect. 3.6.

© Springer International Publishing AG 2017

M.W. Ulmer, *Approximate Dynamic Programming for Dynamic Vehicle Routing*,
Operations Research/Computer Science Interfaces Series 61,
DOI 10.1007/978-3-319-55511-9_3

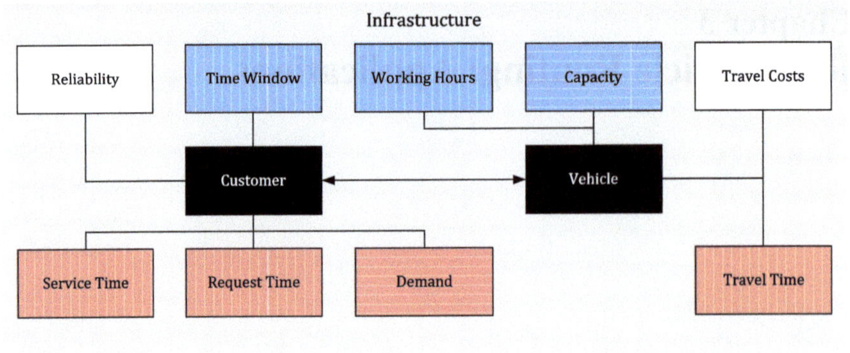

Fig. 3.1 Entities, constraints, uncertainty factors, and objectives of rich vehicle routing applications

We differentiate vehicle routing to conduct services in Sect. 3.7, and to transport passengers and goods in Sect. 3.8. For the different applications, we analyze the requirement for replanning or stepwise planning.

3.1 General RVRP-Entities

In this section, we formally describe the main entities of vehicle routing following Toth and Vigo (2001, p. 1ff). We briefly summarize the most common underlying components based on the literature. These components are highlighted in cursive. Notably, not every vehicle routing application contains all components, but the majority of the applications can be described by a subset of the components.

3.1.1 Infrastructure

The infrastructure is defined by the *service area*. The service area contains a set of *locations*, the *depot* or several depots, and a *traffic network*. The traffic network consists of a set of (street) *segments* connecting the locations. A segment has a certain *status* (e.g., free, blocked, congested, etc.). The segments are combined to *paths*. The paths are assigned to a *travel distance* and *travel time* for every vehicle. Both may depend on the statuses of the segments and the vehicle's *specifications*.

3.1.2 Vehicles

The dispatcher routes a fleet, i.e., a set of *vehicles* (in the traffic network) to conduct services or to deliver or collect goods at *customers*. The vehicles may be heterogeneous and have loading or travel *capacities*. This means that, e.g., the volume of loaded goods and the travel duration or distance are limited. Further, vehicles may have different (driver's) *skills* and *specifications*. The vehicles' specifications may include the *load*, the current *position*, the driver, the current speed, acceleration, maximal speed, technical condition, or emissions. The vehicles may operate during *working hours*. The working hours are defined by an *initial point of time* in which the vehicles start in the depot and a *time limit* by which the vehicles return to the depot at the latest.

In order to serve a customer, a vehicle has to travel to the customer's location. The *path* is the set of connected segments the vehicle uses between origin and destination. The *travel time* between depot and customer or between two customers is the amount of time the vehicle requires between leaving the origin and arriving at the destination. The *travel distance* constitutes the according sum of traveled distances on the segments the vehicle uses between origin and destination. Travel distance and travel time are assigned to *travel costs*.

3.1.3 Customers

Customers located in the service area request the visit of a vehicle for service or transportation. A *request* consists of the point of *request time*, the customer's *location* in the service area, and the requested good or service. To serve a customer, the *demand*, the *service time*, the *time window*, some *requirements* for the vehicle's or driver's skills, the *priority* of the request, a *destination* (e.g., for a cab), the *reward* of delivering the product and the *penalty* of rejecting the request or violating the customer's constraints may be considered. The demand is the amount of goods the customer needs. The service time is the required time to handle the good or to conduct a service, defined by the time span between vehicle's arrival and departure. The service time may depend on the vehicle. The time window of a customer is defined by the earliest and latest possible point of time of the vehicle's visit. The requirement defines the types of vehicle able to fulfill the request. Priority defines the relative urgency of the product to be delivered. Some priorities may allow the rejection of a request. Delivery or pickup of goods and conducting a service result in a *reward*. Violations of time windows or rejections result in *penalties*. Rewards and penalties reflect customer satisfaction and can be combined to the measure *reliability*.

3.2 Plans

Dispatchers, vehicles, and drivers perform *tasks*. Tasks mainly include the delivery or pickup of goods and the conduction of a service, but also the communication with the customers (e.g., request confirmations) and the change of vehicles' specifications (e.g., by refueling or replenishment).

Routing is the determination of *routes* for vehicles. For VRPs, a route is defined as a *sequence* of customers connected by paths. As a result, routing includes the *assignments* of customers to vehicles. Notably, assignments include possible rejections of customers if the customer is not assigned to any vehicle. For every vehicle, the sequence of assigned customers is determined. The *duration* of a route is the sum of the individual travel times on the paths between the customers and service times of the tasks at the customers. The *length* of a route is the sum of the travel distances on the paths between the customers. A *plan* is the combination of routing and tasks. A plan is *feasible* if working hours, time windows, vehicles' capacities, and the requirements of the customers are considered.

3.3 Objectives

The objectives for companies conducting vehicle routing are twofold on the operational level. First, the objective may be to fulfill tasks to minimal costs. Second, the objective may be to use given resources (e.g., vehicles and working hours) effectively to achieve a high service level and customer satisfaction. Hence, we differentiate between *costs* and *reliability*.

3.3.1 Costs

On the operational level, costs mainly contain travel time and distance. Travel distance is a main cost factor for routing in rural areas. In urban environments, travel time is the main cost factor on the operational level. In some cases, dispatchers additionally decide about the number of dispatched vehicles. This number impacts the costs in many cases significantly.

3.3.2 Reliability

Reliability implies fast services, the service of a large number of customers, the compliance of time windows, and overall high customer satisfaction. Even though customer satisfaction may be represented by pecuniary factors as well, for many applications like emergency services, costs on the operational level are secondary. Further, in e-commerce, fast and reliable deliveries are mandatory for customer satisfaction enabling subsequent orders. Time windows can be differentiated in "soft" and "hard". The violation of soft time windows results in a penalty while a violation of hard time windows results in waiting or even a decline of service.

3.3.3 *Objective Measures*

Because the planning is conducted repeatedly on daily basis, LSPs often aim on maximizing mean objective values in customer satisfaction or minimizing mean objective costs respectively. Still, in some cases, LSPs may aim on avoiding a wide spread in objective values. As an example, LSPs may aim on serving a similar amount of (requesting) customers each day to achieve a balanced workload (Ulmer and Voß 2016). Further, LSPs may be risk-averse evaluating high objective values lesser than particularly low objective values or LSPs may even only consider worst-case scenarios.

3.4 Constraints

As main constraints, we differentiate between customers' *time windows*, drivers' *working hours*, and vehicles' *capacities*. The constraints may be *hard*, i.e., a violation may not be allowed. In some cases, constraints can be *soft*, i.e., violations are allowed but result in penalties or costs respectively.

3.4.1 *Time Windows*

In many cases, customers have to attend home to receive the service or delivery. Therefore, time windows are communicated. A time window for a customer is defined by the earliest and latest possible point of time the customer can be visited by a vehicle. If a vehicle arrives too early at a customer, the vehicle may have to wait until the beginning of the time window resulting in (unnecessary) travel costs. If it arrives late, the vehicle may not be able to serve the customer at all, or the late arrival may lead to a penalty and a loss in reliability.

3.4.2 *Working Hours*

For many routing applications, the routing is conducted during working hours. The working hours are defined by an initial point of time in which the vehicles start in the depot and the time limit by which the vehicles have to return. Working hours may limit the number of served customers. Not served customers may result in a loss of reliability. To serve all customers, dispatchers may schedule additional vehicles resulting in an increase in travel costs.

3.4.3 *Capacities*

Vehicles often have only a limited capacity to transport passengers or goods. If customers' demands are higher than the loaded amount of goods, this results in customer dissatisfaction and a loss of reliability. Dispatchers may schedule another visit to fulfill the remaining demand resulting in additional travel costs.

3.5 Drivers of Uncertainty

In practical vehicle routing applications, the sources of uncertainty may be manifold. As an example for uncertain specifications, vehicles may break down, changing the maximal speed to zero (Mu et al. 2011). As an example for uncertain capacities, vehicles may have changing travel ranges. This is especially the case for electric vehicles (Ferreira et al. 2013). Further, the requirements of the customers and the according required skills of vehicles and drivers may be uncertain. In the following, we focus on the most common drivers of uncertainty in real-word application based on the findings of Sect. 2.5. The distinguished drivers match the uncertainty factors mainly analyzed in the literature (Psaraftis et al. 2015). We distinguish *travel time*, *service time*, *demands*, and *requests*.

3.5.1 *Travel Times*

Uncertainty in travel time may be a result of spontaneous changes in the statuses of segments and paths respectively. A change of the statuses results from congestion, accidents, or control of the traffic management as described in Sect. 2.5.2. Notably, uncertain travel times have to be differentiated from varying, but known travel times resulting from time-dependent street loads (Malandraki and Dial 1996).

3.5.2 *Service Times*

In many cases, the required service time is not entirely known before the vehicle arrives at the customer or even before the service is finished. The service time may depend on the time for searching a parking space, or, on the requirements of the customer. This may be especially the case for technicians and maintenance workers. Another field of uncertain service times is healthcare. In many cases, the physicians are unaware of the patient's condition before they arrive. As a result, the required amount of time to spend at the patient's home significantly differs.

3.5.3 *Demands*

In some cases, the volume of customer demands are large and the vehicles' loading capacities have to be considered in planning. This may be the case in oil distribution or waste collection. An insufficient amount of goods or insufficient free space may force the vehicles to return to a depot for replenishment or unloading. In other cases, dispatchers have to maintain a certain *inventory* at the customers. This may be the case if the customers require resources for (flexible) production or in shared mobility systems, where customers spontaneously demand a free vehicle to rent or a free parking spot to return the vehicle.

3.5.4 *Requests*

For many routing applications, the customers requesting for a service, pickup or delivery are unknown before they request, i.e., before the request time. This may be the case for emergency vehicles, technicians, courier services, and passenger transportation. A customer request is described by the customer, the customer's location, the point of time, and the requested service or good.

3.6 Classification

The Cartesian product of objectives, constraints, and uncertainty drivers generates the set of all possible settings in practical routing applications. We use the presented objectives, constraints, and drivers of uncertainty to classify the subset of most prominent practical routing applications based on the literature. Arguably, a sharp distinction is not always possible. We utilize the depictions as a support for the following overview. For the different problems, we analyze the requirement for replanning to reveal SDVRP-applications. In many cases, adaptions of planning or the generation of entire new plans are necessary. We differentiate the application fields regarding requirement for replanning and the according planning horizon as depicted in Table 3.1.

We classify the applications by estimating the length of time for which the dispatcher's plan and routing decisions can be conducted without the requirement for further decision making. We differentiate applications with *long*, *medium*, and *short* time. The requirement for replanning directly induces the frequency of decisions to be made. It is correlated to the possibility to establish and conduct an a priori plan and the severity of the required adaptions.

The requirement for replanning or stepwise planning may be twofold. Planning for the application may be too complex to achieve an a priori plan. In this case, sequential updates of the plan may allow to focus on the relevant information for

Table 3.1 Subsequent planning

Planning	Decision frequency	A priori	Adaptions	Uncertainty
Long	Low	Straightforward	Minor	Slight
Medium	Intermediate	Possible	Major	Significant
Short	High	Impossible	Full	Strong

the specific point of time (e.g., on a rolling horizon). For example, this could be experienced if the travel time changes over the day according to a known variation in the load of the street capacity (Malandraki and Dial 1996). Nevertheless, replanning is mainly necessary if information changes or a percentage of information is only revealed over time, i.e., if the application contains uncertainty. In this case, adaptions of plans or stepwise planning are required to react to the changed circumstances and updated information.

Some applications allow the straightforward determination of a (robust) a priori plan with only a few adaptions during the day. These applications enable planning a relatively long time ahead. Some applications may allow the possibility to plan a priori but require flexible plans allowing constant replanning due to new events. The achieved ex post routing may significantly differ from the original a priori plan. Planning is conducted a *medium* time ahead. For some applications, the definition of an a priori plan may not be useful, is even impossible (e.g., ambulances), or the a priori plan is frequently changed substantially. For these applications, dispatchers mainly conduct stepwise planning on a *short*-term basis aiming on the provision of flexibility regarding new events. For replanning and stepwise planning, the tools depicted in Sect. 2.5 are utilized. We assume that the planning is highly influenced by the uncertainty's impact. Applications with only a slight uncertainty impact may allow planning relatively long-term and robust a priori plans. For applications, where the uncertainty impact is strong, long-term planning may be rendered useless.

3.7 Service Vehicles

With this classification at hand, we now analyze the applications. We start with vehicle routing applications to conduct services depicted in the upper part of Table 3.2. Service vehicles generally do not have any transportation purposes. Application fields of service vehicle routing are emergency, maintenance, construction, repair, healthcare, and salesperson services. Service vehicles may be ambulances, police and fire brigades, healthcare, breakdown, technician, or repair vans. Considering a vehicle fleet, the service of a customer is usually not bound to a specific vehicle but can be reassigned if necessary.

Table 3.2 Vehicle routing in practice

Application	Objective		Constraints			Uncertainty				Planning
	Costs	Reliability	Time windows	Working hours	Capacity	Travel time	Service time	Demands	Requests	
Service										
Emergency		✓	✓		✓		✓		✓	Short
Technician	✓	✓	✓	✓		✓	✓		✓	Medium
Healthcare	✓	✓	✓	✓		✓	✓		✓	Medium
Salesperson		✓	✓	✓			✓			Medium
Passenger transport										
Bus		✓	✓		✓	✓		✓		Long
Dial a ride	✓	✓	✓		✓	✓			✓	Short
Patient transport		✓	✓	✓	✓	✓	✓		✓	Medium
Shared mobility		✓	✓		✓			✓	✓	Short
Goods transport										
Distribution	✓				✓			✓		Long
Waste	✓				✓			✓		Long
Retail	✓	✓	✓	✓	✓	✓	✓			Medium
Grocery	✓	✓	✓	✓		✓			✓	Medium
Parcel delivery	✓			✓		✓				Long
Parcel pickup	✓			✓		✓			✓	Short
Courier services	✓	✓	✓			✓			✓	Short
Same-day delivery	✓			✓		✓			✓	Short

Responses to emergency calls have to be immediate (Schmid 2012). The resulting costs are secondary. Emergency vehicles have priority in traffic and therefore are influenced by the traffic situation only peripherally. Emergency vehicles experience a high degree of uncertainty. Requests and the time span in which the assigned emergency vehicles will not be available are hereby generally unknown (Beraldi and Bruni 2009). A priori planning is hardly possible. Stepwise planning has to be conducted.

For *technician* and *healthcare* services, service providers often arrange a time window (Souyris et al. 2013). Within this time window, the customer attends home to allow the conduction of the service. Technical and healthcare service providers aim on two objectives: On the one hand, customer satisfaction is required to avoid contractual penalties and to achieve subsequent orders. On the other hand, skilled technicians and physicians are expensive to employ. Hence, service providers aim on reducing the daily travel times to conduct more services within the working hours.

For service vehicles in the urban environment, travel times between the customers are often uncertain due to possible congestion and limited road capacities. Service vehicles usually spend a significant amount of time at the customer. In some cases, the service to conduct is itself uncertain (Yuan et al. 2015). Many service calls arrive spontaneously during the day. Customers are often unknown (Chen et al. 2015). As a result, uncertain elements for service vehicles are mainly the travel time between customers, the required service time, and the customer requests during the day.

For technician and healthcare applications, often an a priori plan is given. The frequency of plan adaptions differs. Requests for key services and breakdown vans are merely unknown and adaptions have to be conducted frequently. Stepwise planning is required. For telephone technicians, all customers may be known beforehand. A priori planning is possible. A replanning may only be necessary because of service and travel times.

Salespersons visit clients to sell goods (e.g., insurances, drugs, books, or tools) during the day. They aim on appointments in the visiting hours of the client. In many cases, salespersons are not the clients' top priority and they have to wait (Zhang et al. 2015a). The waiting time or service time respectively is therefore unknown and impacts future appointments. Salespersons may have to reschedule or skip planned appointments.

3.8 Transportation Vehicles

The main purpose of transportation vehicles is to transport goods or passengers from one location to another. An extensive classification of real-world applications for transportation routing is provided by Coelho et al. (2015). In the following, we differentiate between the transportation of passengers and goods as depicted in the lower part of Table 3.2.

3.8.1 Passenger Transportation

In passenger transport, application fields are vast. Passenger transportation mainly consists of public transportation (e.g., buses), demand responsive passenger transport like dial a ride, cabs, patient transportation, and shared mobility systems.

Road-based public transportation is generally performed by *buses*. These buses follow regular a priori schedules generated on the tactical planning level. The routing is generally not alterable. For generation of robust schedules, dispatchers have to consider the operational uncertainty and constraints in routing. Since bus lines are often located in an urban environment, schedules have to consider varying travel times. Further, because the capacity of a bus is limited, demands have to be included in planning. Bus services are an example of a routing application, where all planning is already conducted while creating the schedule. As a result, the a priori plan holds. Adaptions are generally not necessary.

Dial a ride and cabs are usually called spontaneously to pick up a customer at an origin location and drop him or her off at a destination location (Häll et al. 2012). They have to work both reliably and efficiently. In many cases, the customer has to arrive at the destination (e.g., airport, train station) before a deadline. Dispatchers aim on serving many customers per hour minimizing empty tours and the driver's idle time. To hold the deadlines, dispatchers have to consider uncertain travel times (Schilde et al. 2014). The requests for cabs and dial a ride services are generally unknown beforehand. As a result, dispatchers experience uncertainty in travel times and customer requests. An a priori planning for dial a ride is generally not possible.

Dispatchers only can plan stepwise aiming on flexibility with respect to new requests. For every new request, assignment and sequencing decisions have to be determined.

In the transportation of *patients*, e.g., elderly or disabled persons, customers are picked up at their homes and are dropped off at, e.g., a hospital or a physiotherapy (Zhang et al. 2015b). Later, the customers are driven back to their homes. To match the customers' appointments, dispatchers have to consider time windows (Ritzinger et al. 2014). Further, the customers have to return to their homes within the drivers' working hours. Finally, vehicles can only transport a limited number of customers per time. To satisfy the time windows, dispatchers have to consider uncertain travel times within the city. Since many customers need assistance when entering the vehicle, a certain amount of service time is required at the customers' homes and the destinations (Beaudry et al. 2010). In many cases, this service time is uncertain. Further, the point of time in which the customer requests to be picked up to be brought back home may be uncertain as well. As a result, dispatchers have in many cases to replan the assignments and customer sequences. The resulting applications generally allow a priori planning but require adaptions of the plan due to the manifold sources of uncertainty.

Shared mobility systems enable customers to spontaneously rent a bike or car within the city or at specific stations. The main objective of operational planning for shared mobility systems is to provide a sufficient service level (Lin and Yang 2011). This means that a customer can rent a vehicle at any location at any point of time. Especially bike sharing systems are highly volatile. The requests are ad hoc and may occur at any station in the whole urban area. At every station, an inventory of bikes is given. This inventory varies due to customer rentals, i.e., requests. The number of requests, i.e., the demand for bikes is unknown. To provide a sufficient inventory, i.e., number of available bikes and free bike racks at any station and at any point of time, dispatchers route a set of capacitated vehicles picking up and returning bikes between stations. Time windows indicate in this case the necessity to deliver bikes at a certain point of time to fulfill a request. Since the realized requests change the inventories at the station, a constant adaption of the plan is required and an a priori plan might not be effective (Brinkmann et al. 2015).

3.8.2 Transportation of Goods

In the transportation of goods, we differentiate between distribution of homogenous goods (e.g., oil), waste collection, retail delivery (and assembly), grocery delivery, parcel delivery, parcel pickup, courier services, and same-day parcel delivery.

Given the *distribution* of a homogenous good like heating oil, the vehicle serves a predefined sequence of customers (Novoa and Storer 2009). Since the vehicle is capacitated, the vehicle has to return to the depot for replenishment. The exact demands of the customers are often unknown. Dispatchers aim on efficient depot returns considering the uncertain demands. The same behavior can be experienced for the collection of waste. Again, the amount of waste at a customer is unknown.

Therefore, vehicles may return to the depot for emptying. Both applications show a low requirement for adaptions. Except for the depot revisits, the routing is mainly predefined and not changed during the day.

Retail delivery contains delivery and in many cases assembly or connection of a certain good, e.g., a dishwasher. The delivered goods are usually large resulting in *less than truckload* routing. For delivery and installation, time windows are communicated. Dispatchers aim on satisfying the time windows cost efficiently. They have to consider uncertain service times for installation and uncertain travel times. In contrast to service vehicles, customer assignments are unalterable because the good is loaded to a specific vehicle. Hence, time windows have to be updated or deliveries have to be postponed in some cases. This results in occasional adaptions of the a priori plan.

Many supermarkets now offer the delivery of *groceries* to the customers. Customers can order online and determine a time window for delivery. Providers collect these orders in a collection phase and deliver the goods in a subsequent delivery phase. Customers have to attend home during this time window to receive the goods. To satisfy the time windows, dispatchers have to route vehicles considering uncertain travel times. Even though requests occur before the execution of the routing, dispatchers have to consider uncertain requests in the assignment of the time windows (Ehmke and Campbell 2014). Dispatchers aim on efficient delivery. Simultaneously, they need to consider customer satisfaction because late deliveries may result in the loss of customers and subsequent orders (Cleophas and Ehmke 2014).

The field of parcel pickup and delivery is vast. We differentiate between plain parcel delivery and parcel pickup, combined pickup and delivery often conducted by courier services, and the emerging field of same-day delivery. In *parcel delivery*, the vehicles start at the depot loaded with a set of parcels. These parcels are delivered to the customers over the day following an a priori plan. The objective is to conduct these deliveries efficiently. Dispatchers usually have to consider varying travel times. The requirement for adaptions of the plan is rarely given (Groß et al. 2015).

Especially e-commerce companies often request *parcel pickups* to allow a fast dispatching the same day. These pickup-requests occur during the day and are unknown beforehand. Dispatchers route vehicles to collect the parcels and to deliver them to the depot for long haul shipping (Pillac et al. 2013). If some pickup-requests are known in the beginning, an a priori routing can be planned. Nevertheless, an adaption of the plan is necessary for every new request. As a result, the a priori plan may be inefficient and stepwise planning may be suitable.

Courier services are similar to parcel pickups but often pick up goods at one location in the city and instantly have to deliver them to another location in the city (Berbeglia et al. 2010). Both locations are unknown before the request occurs. Courier services often transport urgent goods like blood or documents. Therefore, the delivery has to be conducted before a deadline. For courier service, dispatchers are not able to plan an a priori routing, but have to react to every new request.

The field of *same-day delivery* is still emerging. The main goal is to cost efficiently deliver the good the same day the order was placed. The goods may be stored at the depot or have to be picked up at a store. Mainly, orders are accumulated in a collection phase and distributed during a delivery phase within working hours (Slater 2002). Recently, these phases start to overlap meaning that orders occur when the vehicles are already delivering goods (Voccia et al. 2017). As a result, the vehicles have to be rescheduled for every new order.

3.9 Implications

As presented in this chapter, the application fields requiring vehicle routing with respect to uncertainty and constant adaptions of the plans are vast. We can identify two application classes regarding uncertainty and replanning. For the first class, a (robust) a priori plan can be conducted with occasional adaptions. For the first application class, (static) modeling as proposed in Toth and Vigo (2001) may be suitable. Uncertainty may be transformed into deterministic parameters using preprocessing. The planning can then be conducted by well-established deterministic methods (Chardy and Klopfenstein 2012). In the following, we will not further consider this class. In the remainder of this book, we focus on the second application class, indicated by *medium* or *short* time spans between replanning situations. This class of applications requires significant changes of the a priori plan or the determination of an a priori plan is not possible at all. Plans have to flexible regarding future uncertain events. Even though both classes can be described as SDVRPs, we denote the second class as SDVRPs in the narrower sense. We now briefly describe the required steps to achieve anticipatory quantitative decision support for this class.

3.9.1 Decision Support

Service providers operate on a highly competitive market with low profit margins. For economical success, vehicle routing plans have to be both efficient and effective. Especially for the second problem class, the determination of suitable plans is challenging and the requirement for scientific quantitative decision support tools is high. To allow quantitative decision support for vehicle routing, the presented problems have to be scientifically modeled. For SDVRPs, a classical static modeling is not suitable. The model has to reflect both uncertainty and replanning to allow the application of prescriptive analytics and the development of suitable solution algorithms and flexible plans. The scientific models of the according real-world components are depicted in Table 3.3. Especially for applications requiring constant replanning, *anticipation*, i.e., an integration of possible uncertain future events and future decisions in the current planning may avoid myopic, ineffective decisions and maintain flexibility. As a result, applied solution approaches need to allow anticipation. In the following, we present the required steps to model real-world applications and to achieve anticipatory quantitative solution approaches.

Table 3.3 Modeling steps

Real-world	Model
Planning situation	Decision state
Uncertainty	Stochasticity
Stepwise planning	Dynamism
SDVRP	MDP

3.9.2 Modeling of Planning Situations

First, we require a scientific model of a dispatcher's planning situation in context of the vehicle routing components described in Sect. 3.1. In Sect. 4.4, we model a planning situation as a *decision state*. This model allows us to achieve unambiguous definitions and quantification of planning situations for the main applications presented in this chapter.

3.9.3 Modeling of Uncertainty

Uncertainty in real-world applications has to be integrated in problem models to apply suitable solution approaches. Uncertainty is therefore modeled as *stochasticity* by methods of predictive analytics. As an example, the uncertain travel time for a path is modeled as a stochastic probability distribution. The properties of the distribution to resemble the behavior of the observed real-world travel times can, e.g., be derived from predictions or historical observations. The process of modeling and the general way to model the four main drivers of uncertainty are shown in Sect. 4.5.

3.9.4 Modeling of Subsequent Planning

For stepwise planning and constant adaptions of the plans, the model must reflect the characteristics of the decision. As an example, an a priori plan may define the sequence of customers to visit. Nevertheless, in stepwise planning, only the decision where to go next has to be made. To achieve effective decisions, decision making needs to anticipate following decision points. The model has to reflect the possibility of replanning during the day or to plan stepwise. The possibility of subsequent planning is generally modeled by *dynamic* decision problems recalled in Sect. 4.1.

3.9.5 Modeling of Applications

The three vehicle routing (sub-)models of planning situations, uncertainty, and subsequent planning have to be combined to an overall model of the SDVRP. Further, both an overall objective and an objective at a specific decision point have to be determined. In Sect. 4.2, we recall the (finite) MDP as the canonical model for defining dynamic and stochastic vehicle routing problems allowing the consideration of stochasticity and highlighting the significant decision factors in every decision point. In Sect. 4.6, we show how SDVRPs are modeled as MDPs.

3.9.6 Modeling of Anticipation

Future uncertain events in the real world are usually depicted by forecasts based on predictions or the analysis of historical data as described in Sect. 2.5.5. These forecasts have to be integrated in the model and solution method. Solution methods are able to utilize the regarding MDP quantities to allow the inclusion of future decision making and stochasticity in the derivation of a decision policy. This inclusion is called *anticipation*. Anticipation is mandatory for the problem class with short or medium planning. Anticipation can be achieved in several ways. In Chap. 5, we define and classify anticipation regarding the MDP.

3.9.7 Anticipatory Methods

To achieve anticipation in sequencing and assignment decisions, suitable quantitative solution methods for SDVRPs from the field of prescriptive analytics are required. In Chap. 6, we give an overview of anticipatory solution methods regarding the classification derived in Chap. 5. Especially for the second class of presented applications, uncertain (stochastic) information has to be included to avoid myopic decision making. Hence, we focus on methods allowing the explicit integration of uncertainty, i.e., stochastic information in the solution approach.

Chapter 4
Modeling

To apply quantitative solution algorithms to the applications presented in Chap. 3, the applications at hand have to be modeled rigorously. A model covers "all relatively stable and general features of the world that are interesting from a scientific point of view" (Frigg and Hartmann 2012). A model allows a (simplified,) unambiguous definition and quantification of a real-world system. Strictly speaking, the formal depiction of the entities in Sect. 3.1 is already part of the modeling.

In this chapter, we model vehicle routing problems containing uncertainty and requiring stepwise planning by a (finite) Markov decision process as proposed by Powell et al. (2012). For modelling, an MDP is suitable because it allows to model uncertainty and stepwise planning explicitly. We define the MDP in Sect. 4.2. The MDP contains three (sub-)models: decision states, dynamic decision making, and stochastic transitions. To model routing applications as MDP, we have to determine the three (sub-)models. In Sect. 4.1, we model replanning as *dynamism*. In Sect. 4.4, we model a *decision state* for a vehicle routing's planning situation regarding the entities and constraints described in Sect. 3.1. We model uncertainty as *stochasticity*. We give a short overview on how the main drivers of uncertainty generally are modeled in the literature and derive stochastic transitions for the MDP in Sect. 4.5. We then give an overview on how SDVRPs are modeled as MDPs in Sect. 4.6.

4.1 Stochastic Dynamic Decision Problem

In this section, we describe dynamic decision making based on a dynamic decision problem. First, we need to formally define the terms *stochasticity* and *dynamism*.

© Springer International Publishing AG 2017

M.W. Ulmer, *Approximate Dynamic Programming for Dynamic Vehicle Routing*,
Operations Research/Computer Science Interfaces Series 61,
DOI 10.1007/978-3-319-55511-9_4

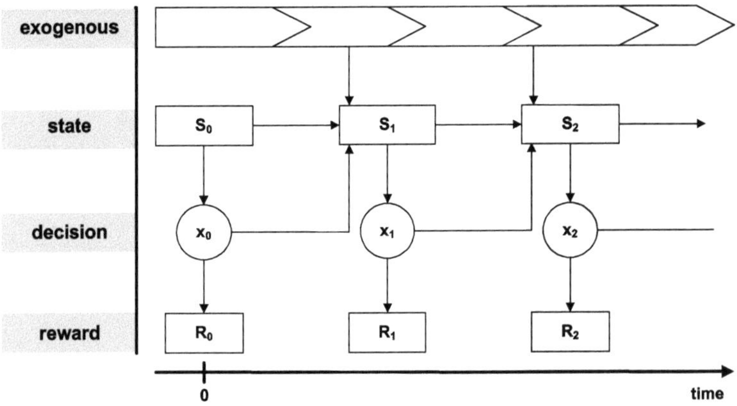

Fig. 4.1 Dynamic decision process (Meisel 2011)

If a problem only consists of a single decision point, it is called *static*. If it contains a sequence of decision points, it is called *dynamic*. A decision problem where all information is known at every time is called *deterministic*. A decision problem where some information is only revealed over time is called *stochastic*. Since deterministic problems only require a single decision, only stochastic problems are dynamic (Lahyani et al. 2015).

Notably, some authors define (stochastic) problems where new information reveals over time as dynamic and deterministic if no knowledge about the problem's stochasticity is given, e.g., Bektas et al. (2014). We do not share this opinion.

4.1.1 Dynamic Decision Problems

A dynamic decision problem contains of an exogenous process of the *object system*. The synchronization with the planning process proves *states* at certain points of time. For each state, the selection of applied actions is required. These selections are called *decisions*. Each decision provides a certain outcome, the so called *reward*. The decision and the exogenous process's impact lead to a *transition* resulting in the next *decision point*. If the impact of the exogenous process is not certain, the dynamic decision problem is stochastic.

The process of dynamic decision making is presented in Fig. 4.1. Here, states are denoted by S_i, decisions by x_i, and rewards by R_i. Decision points occur over a time horizon, but are not necessarily assigned to certain times. In many cases, the applied decision and the object system influence the time of the next decision point.

4.2 Markov Decision Process

The canonical way to model a stochastic dynamic decision problem is as a Markov decision process (Bellman 1956). Notably, modeling is independent of the applied solution method. Nevertheless, an MDP model may facilitate the derivation of anticipatory approaches. In this section, the definition of MDP is given in Sect. 4.2.1. We only consider finite and discrete MDPs to describe stochastic and dynamic vehicle routing problems. For an extensive overview on MDPs, the interested reader is referred to Puterman (2014). Decision policies and realizations within an MDP are defined in Sect. 4.2.2.

4.2.1 Definition

A stochastic dynamic decision problem can be defined using a Markov decision process. Within the MDP, a number of decision points $\mathscr{K} = \{0, \ldots, K-1\}$ occurs subsequently. Here, K can be a random variable. For each decision point $k \in \mathscr{K}$, a set of states $\mathscr{S}_k$ is given, combined in the finite set of states $\mathscr{S}$ as denoted in Eq. 4.1. $\mathscr{S}$ is called the *state space*.

$$\mathscr{S} = \bigcup_{k \in \{0,\ldots,K\}} \mathscr{S}_k \tag{4.1}$$

$S_0 \in \mathscr{S}$ denotes the *initial state* and $S_K \in \mathscr{S}$ denotes the *termination state*. For each decision point $k \in \mathscr{K}$ and for each state $S_k \in \mathscr{S}$, a subset of decisions $\mathscr{X}(S_k) \subseteq \mathscr{X} = \{x_1, \ldots, x_m\}$ of the overall set of decisions $\mathscr{X}$ is given. $\mathscr{X}$ is called the *decision space*. The combination of a state S_k and a decision x leads to a post-decision state (PDS) $(S_k, x) = S_k^x \in \mathscr{P}_k \subseteq \mathscr{P} = \mathscr{S} \times \mathscr{X}$, with $\mathscr{P}$ the overall set of post-decision states (Van Roy et al. 1997). $\mathscr{P}$ is called the *post-decision state space*. The post-decision state is the deterministic result of the combination between state and decision. Notably, different tuples of states and decisions may lead to the same PDS: $S_k^x = (S_k, x_i) = (\hat{S}_k, x_j)$ with $\hat{S}_k \neq S_k$ and $x_i \neq x_j$. The application of decision x given state S_k leads to an immediate reward (or costs) $R(S_k, x)$ with $R: \mathscr{S} \times \mathscr{X} \to \mathbb{R}$. Given PDS S_k^x, a stochastic transition $\omega_k \in \Omega_k$ with $\Omega_k \subset \mathscr{P} \times \mathscr{S}$ leads to the next state $(S_k^x, \omega_k) = S_{k+1} \in \mathscr{S}$. The probability of a transition $\omega_k = (S_k^x, S_{k+1}) \in \Omega_k$ between PDS $S_k^x \in \mathscr{P}$ and state $S_k \in \mathscr{S}$ is given by $\mathbb{P}: \Omega_k \to [0, 1]$. $\mathbb{P}$ holds $\sum_{S \in \mathscr{S}} \mathbb{P}(\omega_k = (S^x, S)) = 1 \; \forall \, S^x \in \mathscr{P}$. Notably, a deterministic and static decision problem can be modeled as the special case of an MDP, where $\mathbb{P}(S_k^x, S_{k+1}) \in \{0, 1\}$ and $K = 1$. For an extensive overview of dynamic programming, the interested reader is referred to Bellman (1956).

For each decision point, the MDP contains of a decision component. The whole MDP is the composition of all components. Figure 4.2 shows the functionality of a decision component. In decision point $k \in \mathscr{K}$, a current system state $S_k \in \mathscr{S}$ and a

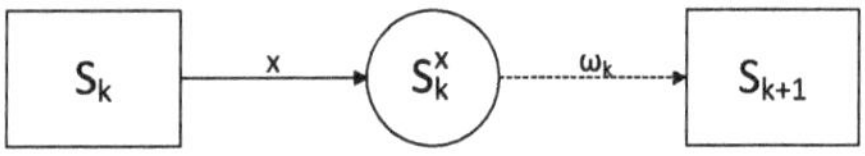

Fig. 4.2 Component of a Markov decision process

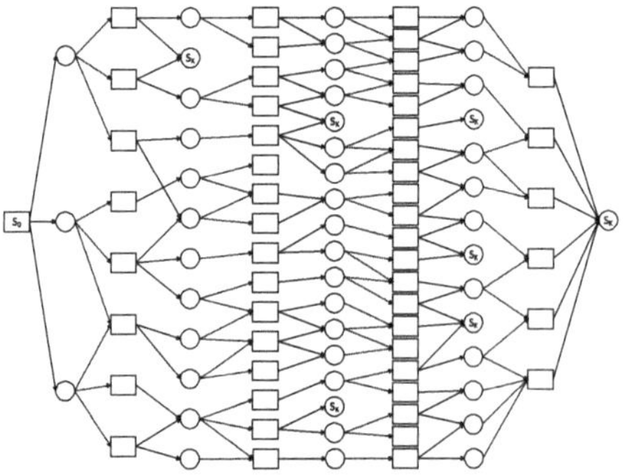

Fig. 4.3 Example of a Markov decision tree

set of feasible decisions $\{x_1, \ldots, x_l\} = \mathscr{X}(S_k) \subseteq \mathscr{X}$ dependent on state S_k are given. The application of a decision $x \in \mathscr{X}(S_k)$ leads to an immediate reward $R(S_k, x)$ and a post-decision state $S_k^x \in \mathscr{P}$. The stochastic transition $\omega_k = (S_k^x, S_{k+1})$ leads to the next state S_{k+1}.

A Markov decision tree now contains all possible states, decisions, PDSs, and transitions and is therefore a composition of components as presented in Fig. 4.2. Figure 4.3 shows an exemplary Markov decision tree. The circles represent PDSs $S^x \in \mathscr{P}$, the squares states $S \in \mathscr{S}$. The solid lines indicate decisions $x \in \mathscr{X}$, the dashed lines stochastic transitions $\omega_k \in \Omega_k$. The MDP starts on the left, with initial (decision) state S_0. Termination (post-decision) state is S_K. In this case, the termination occurs after the application of a decision. For some problems, termination occurs after a stochastic transition. For the purpose of presentation, we depict the termination state multiple times in Fig. 4.3. The problem depicted in Fig. 4.3 contains at most $K = 5$ decision points, $|\mathscr{S}| = 52$ states, $|\mathscr{P}| = 42$, $|\mathscr{X}| = 75$ decisions and according rewards, and 70 transitions.

4.2.2 *Decision Policies and Problem Realizations*

Decision policies determine the decision to be selected given a specific state. A decision policy $\pi \in \Pi$ is a sequence of decision rules $(X_0^\pi, X_1^\pi, \ldots, X_{K-1}^\pi)$ for every decision point $k \in \mathscr{K}$. Each decision rule $X_k^\pi(S_k)$ specifies the decision x_k^π to be selected when the process occupies state S_k. Referring to Fig. 4.3, for each state, a decision policy removes all decision lines except for one. For the presented decision tree, the number of different policies is $|\Pi| = 3^3 \cdot 2^{20} = 28,311,552$. At three states, three different decisions are possible. At 20 states, two different decisions can be selected. On the left side of Fig. 4.4, the Markov decision tree for an exemplary decision policy π_e is shown. For each state, all decision branches are removed except for one. The gray areas of the decision tree cannot be reached applying policy π_e.

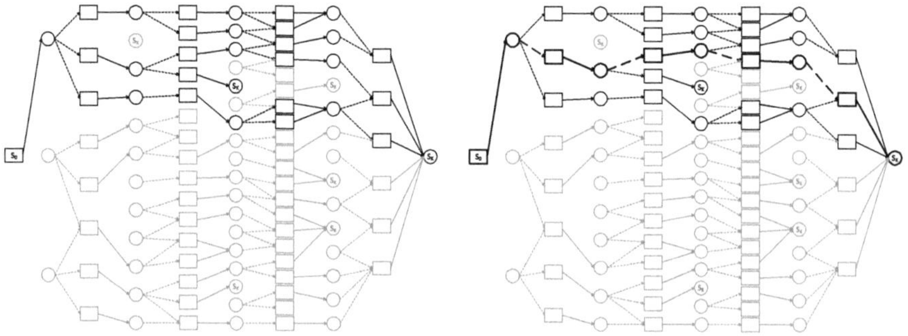

Fig. 4.4 Exemplary decision policy π_e and problem realization ω

A *realization* $\omega = (\omega_0, \ldots, \omega_{k-1}) \in \Omega$ of the set of realizations Ω manifests in the assignment of a transition $\omega_k = (S_k^x, S_{k+1})$ to each post-decision state S_k^x. Referring to Fig. 4.3, for each post-decision state, a realization removes all transition lines except for one. The overall number of realizations for this exemplary decision tree is $3^3 \cdot 2^{19} = 42,467,328$. The number of possible realizations for the exemplary decision policy π_e is $3 \cdot 2^6 = 192$. On the right side of Fig. 4.4, a realization given decision policy π_e is indicated by the bold lines.

4.3 Stochastic Dynamic Vehicle Routing

In the following, we embed SDVRPs in the context of MDPs. First, we recall stochasticity and dynamism in context of vehicle routing. Then, we model a planning situation as an MDP state in Sect. 4.4. We then model uncertainty as stochasticity in Sect. 4.5 and transfer the stochasticity to MDP-transitions or realizations respectively. We finally give examples of SDVRP-models as MDP in Sect. 4.6. To illustrate the differences in static versus dynamic and deterministic versus stochastic vehicle routing, an exemplary problem is presented for each of the three problem classes.[1] All problems are modifications of the classical traveling salesperson problem (TSP). For all problems, the objective is to minimize travel costs. The different problem settings are shown in Table 4.1. An example for a static and deterministic decision problem is the original TSP (Chvátal et al. 2010). The static decision defines the route to travel. All travel costs are known at the time of the decision. An example for a static and stochastic decision problem is the probabilistic TSP (PTSP, Jaillet

[1]Notably, in the presented definition, a deterministic problem is always static. Still, the applied solution approach may be dynamic, e.g., applied on a rolling horizon.

Table 4.1 Classes of TSPs

	Static	Dynamic
Deterministic	TSP	n/a
Stochastic	PTSP	DTSP

1988). The tour has to be defined a priori, but some customers do not require service and are not visited in the actual tour. Dynamism in vehicle routing is linked with stochasticity and often, but not always, with a temporal problem dimension. An example for a stochastic and dynamic decision problem is the dynamic TSP (DTSP, Psaraftis 1988). In the beginning, a subset of customers have to be visited. During the execution of the route, further customers require service and have to be visited. The routing can be adapted to include the additional customers in the tour. For this problem, an optimal solution is not a set of routes, but a decision policy, determining the decision in every possible state. A counter-example for a (deterministic) routing problem with temporal dimension is the time-dependent TSP (TDTSP, Gouveia and Voß 1995). The problem is static, since an optimal solution is a set of routes and can be defined a priori.

4.4 Modeling Planning Situations

In this section, we model the real-world planning situation as a problem's *decision state*. We further model the according decisions and post-decision states.

4.4.1 Decision State

A state S contains every entity described in Sect. 3.1 and an according quantification. In the following, we present how the entities are modeled. Therefore, we (mathematically) model all entities as *elements* of the state. Notably, the models of different SDVRPs may not require all presented elements or may require additional elements representing some underlying components of the entities as described in Sect. 3.1. For the purpose of presentation, we focus on the presented objectives, entities, constraints, and realizations of uncertainties. In the following, we describe how they can be modeled and quantified. An overview of the entities and the according elements is given in Table 4.2. We assume the planning situation occurs at a certain point of time, modeled as $t \in \mathbb{R}_+$.

The infrastructure contains the service area, the depots, the locations of (potential) customers, and the according travel times and distances. The service area $\mathscr{A}$ is usually modeled as a finite subspace or as a limited subset of the two-dimensional

Table 4.2 Modeling planning situations

Planning situation	Decision state-element
Point of time	t
Service area	$\mathscr{A}$
Location	$l \in \mathscr{A}$
Depots	$\mathscr{D} = \{D_1, \ldots, D_p\}$
Customers	$\mathscr{C} = \{C_1, \ldots, C_n\} \subset \mathscr{A}$
Vehicles	$\mathscr{V} = \{V_1, \ldots, V_k\}$
Travel distance of l_1, l_2	$d(l_1, l_2) \in \mathbb{R}$
Travel time of l_1, l_2	$\tau(l_1, l_2) \in \mathbb{R}$
Location of C_i	$l_i^c \in \mathscr{A}$
Time window of C_i	$(t_i^{\text{first}}, t_i^{\text{last}})$
Service time of C_i	$t_i^{\text{service}} \in \mathbb{R}$
Demand of C_i	$c_i^{\text{demand}} \in \mathbb{R}$
Status of C_i	$r(C_i) \in \mathbb{N}$
Travel distance matrix	$\mathscr{M}^d$
Travel time matrix	$\mathscr{M}^\tau$
Position of V_i	$l_i^v \in \mathscr{A}$
Loading capacity of V_i	$c_i^{\text{load}} \in \mathbb{R}$
Fill level of V_i	$c_i^{\text{fill}} \in \mathbb{R}$
Status of V_i	$v_i \in \mathbb{N}$
Working hours of V_i	$(t_i^{\text{init}}, t_i^{\text{limit}}) \in \mathbb{R}^2$
Routing	$\mathfrak{R}$
Route	Θ
Duration	$\bar{\tau}(\Theta) \in \mathbb{R}$
Length	$\bar{d}(\Theta) \in \mathbb{R}$

real plane: $\mathscr{A} \subset \mathbb{R}^2$. The locations l are elements of the service area: $l \in \mathscr{A}$. The depots $\{D_1, \ldots, D_p\} = \mathscr{D}$ are a subset of the service area: $\mathscr{D} \subset \mathscr{A}$.

The travel distance between two locations $l_1, l_2 \in \mathscr{A}$ is defined by $d(l_1, l_2) \to \mathbb{R}$. The travel time is defined by $\tau(l_1, l_2) \to \mathbb{R}$. Usually, travel distances and times are positive. We assume there exist an unambiguous connection between two locations $l_1, l_2 \in \mathscr{A}$. Notably, this may be different considering varying travel times on a street network.

Customers are defined by the set $\mathscr{C} = \{C_1, \ldots, C_n\}$. A customer C_i is represented by the location $l_i^c \in \mathscr{A}$, the time window $(t_i^{\text{first}}, t_i^{\text{last}}) \in \mathbb{R}^2$, the service time $t_i^{\text{service}} \in \mathbb{R}$, and the demand $c_i^{\text{demand}} \in \mathbb{R}$. Further, customer C_i is assigned to a status $r(C_i) \in \mathbb{N}$. This status may represent if a customer is already served (Thomas 2007). The travel distance matrix and travel time matrix of $\mathscr{C}$ and $\mathscr{D}$ are $\mathscr{M}^d$ and $\mathscr{M}^\tau$.

Vehicles are modeled as the set $\mathscr{V} = \{V_1, \ldots, V_h\}$. The current (or target) position of V_i is location $l_i^v \in \mathscr{A}$. A vehicle is further represented by the working hours $(t_i^{\text{init}}, t_i^{\text{limit}}) \in \mathbb{R}^2$, loading capacity $c_i^{\text{load}} \in \mathbb{R}$, and the current fill level $c_i^{\text{fill}} \in \mathbb{R}$. Every vehicle V_i is assigned to a status $v_i \in \mathbb{N}$. For example, this status may indicate if a vehicle is currently occupied or free for service.

4.4.2 Decision Making

Given a (pre-decision) state, dispatchers have to set of possible decisions to conduct resulting in costs or rewards and the post-decision state.

Decisions $x_i \in \mathscr{X}$ mainly contain the movement of vehicles. For every unoccupied vehicle, the next customer to be visited has to be selected. Vehicles may also wait at certain locations. For some problems, decisions are also made about the tasks to conduct. Tasks are highly dependent on the actual vehicle routing problem. A task may include, e.g., the delivery of a good, the interactions between vehicles or the confirmation or rejection of customer requests.

The rewards or costs $R(S, x)$ of a decision x given state S are again highly problem-specific. Rewards may be achieved, e.g., by delivering a good, or serving or confirming a customer request. Costs can occur by traveling or by the violation of a time window resulting from the decision.

The application of a decision x to a pre-decision state S results in a post-decision state S^x. A PDS consists of a subset of elements of S. The exact differences between elements of PDS and state are again highly problem-specific. Usually, the statuses of vehicles change from free to occupied.

To check feasibility of a decision, i.e., a post-decision state, the *hard* constraints have to be satisfied in the current point of time and in the future. Therefore, a feasible plan of the future has to exists. Notably, this plan is not necessary to describe pre-decision and post-decision states. It is rather a tool to ensure feasibility and may be used as basis for planning in the next planning steps.

A decision is called *feasible* if there exist at least one plan for the remaining decision points holding the constraints. A plan contains the routing of the vehicles and the according planned tasks. Routing $\mathfrak{R}$ is modeled as the assignment of customers to routes for every vehicle $\mathfrak{R} = \{\Theta_1, \ldots, \Theta_k\}$. A route Θ_i consist of an ordered subset (sequence) of customers $\Theta_i = (C_{i_1}, \ldots, C_{i_l} = D) \subset \mathscr{C} \cup \mathscr{D}$ usually ending in a depot D. The duration of a route Θ_i is $\bar{\tau}(\Theta_i)$ consisting of the overall travel times between the customers and the overall service times at the customers as depicted in Eq. 4.2.

$$\bar{\tau}(\Theta_i) = \sum_{j=1}^{l-1} \left(\tau(C_{i_j}, C_{i_{j+1}}) + t_{i_j}^{\text{service}} \right) \tag{4.2}$$

The length of Θ_i is $\bar{d}(\Theta_i)$ consisting of the single distances between the customer as depicted in Eq. 4.3.

$$\bar{d}(\Theta_i) = \sum_{j=1}^{l-1} d(C_{i_j}, C_{i_{j+1}}) \tag{4.3}$$

4.5 Modeling Uncertainty

In this section, we present models of uncertainty as stochastic information. The derivation and the modeling are achieved in the field of *predictive analytics*. We focus on the representations mainly used in the literature for the four different drivers of uncertainties. We show that uncertainties are generally modeled as random variables drawing on a (finite) *set* of possible outcomes or a (continuous) probability *function*. The stochastic properties may depend on the point of time. For example, travel times increase during the peak hours. In the following, we model uncertainties for a particular point of time. The time-dimension can be added to the models in a subsequent step.

4.5.1 Deterministic Modeling

In cases where the uncertainty's impact is small, uncertainty can be transferred to deterministic parameters in a preprocessing step (Chardy and Klopfenstein 2012). The deterministic parameters then allow the application of straightforward deterministic solution approaches to achieve a robust a priori solution. This is for example conducted by Vogel and Mattfeld (2011) for stochastic demands in shared mobility. For SDVRPs, deterministic modeling is generally not suitable because the impact of uncertainty is significant and forces dispatchers to replan.

4.5.2 Travel Time

To model uncertain travel times, the travel time τ is modeled as a random variable. In some cases, the statuses of street segments are modeled as a random variable (Van Woensel et al. 2008) but usually, the information is condensed to stochastic travel times per path. Stochastic travel times are mainly represented by stochastic probability distribution. In an urban environment, this distribution is generally assumed to be a Burr, Weibull, Gamma, lognormal, or phase type-distribution because they all enable the properties of a positive skew and long upper tails. An elaborate discussion of suitable distributions is given in Susilawati et al. (2013) and Gómez et al. (2015).

Figure 4.5 shows an exemplary Burr-distribution with a mean value of one. The positive skew is indicated by the major amount of the density being below the mean value. The long upper tail is indicated by the relatively high possible values with low probability. The distribution resembles the actual uncertainty behavior of travel time for a street segment. In the segment is not congested, the travel time does usually not differ significantly. If it is congested, the realized travel time can be significantly higher than the expected travel time.

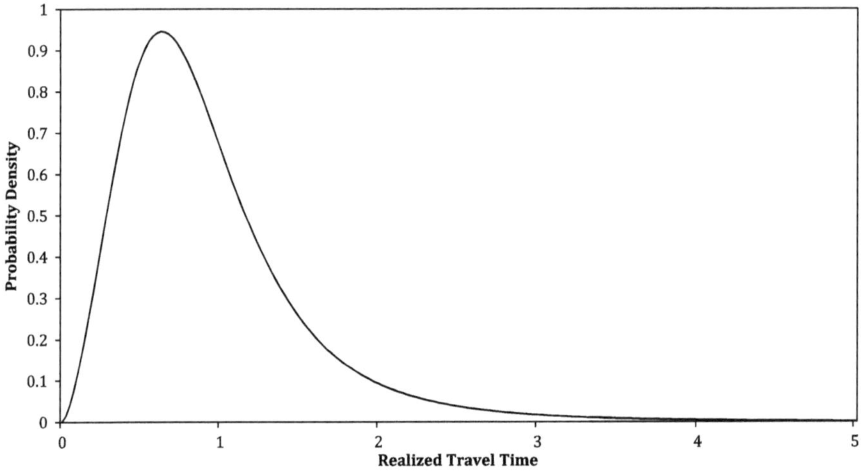

Fig. 4.5 Density function of a Burr-distribution

The main drawback of probability distributions for travel times on a street segment or path between customers is that for a route, i.e., sequence of customers and paths, a combination of several distribution has to be considered. The combination can usually not be represented in closed form, i.e., by a single distribution (Ehmke et al. 2016). An integration in solution approaches is therefore computationally challenging (Gómez et al. 2015). To allow the application of a straightforward algorithm, instead of assuming a probability distribution, travel time is modeled as a set of (equiprobable) real-word observations (Ehmke et al. 2016) or by intervals (Montemanni et al. 2007). The interval range, e.g., is based on the quantiles of the real-world observations (Groß et al. 2015).

In vehicle routing, the travel times on different segments, paths, and between customers are generally considered as *independent*. The realization of one segment does not influence the realization of another. Recently, first work on dependent travel times has been conducted. Köster et al. (2015) model uncertain travel times as a finite set of travel time matrices with uncertain transitions between the matrices. This allows the modeling of travel time dependencies.

4.5.3 Service Time

Stochastic service times $t^{service}$ usually follow probability distributions. These are, e.g., triangular (Errico et al. 2016) or normally distributed (Li et al. 2010). In many cases, service times shows the same behavior as travel times with respect to uncertainty. Hence, Gómez et al. (2015) propose the modeling as phase-type distribution. Mainly, the service time is considered independent of the vehicle. Sungur et al. (2010) model service time as a finite set of possible realizations based on historical observations. Souyris et al. (2013) additionally model the set of realizations dependent on both vehicle (i.e., technician) and customer.

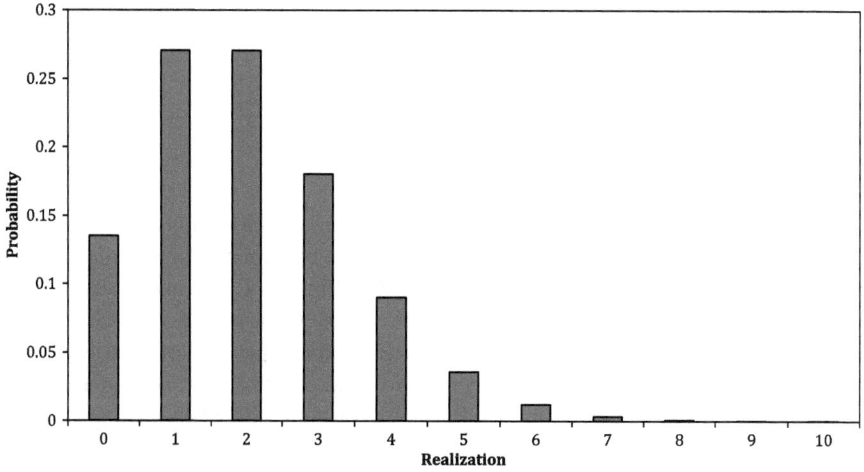

Fig. 4.6 Poisson distribution for $\mu = 2$

4.5.4 Demands

The modeling of demands depends on the application. For distribution and collection applications and inventory routing, demands are modeled by uniform (Secomandi 2000), normal (Coelho et al. 2014), or general unimodal distributions (Goodson et al. 2013a). Secomandi and Margot (2009) model demands by two steps. First, the demand category is determined by *low*, *medium*, and *high*. For each category, the actual demand realization lays within a (discrete) uniform distribution. Erera et al. (2009) model the demands for an inventory routing problem by a set of historical observations.

In shared mobility problems with fixed stations, the demand per customer (i.e., station) is accumulated by a set of trips. Considering only a single station, this results

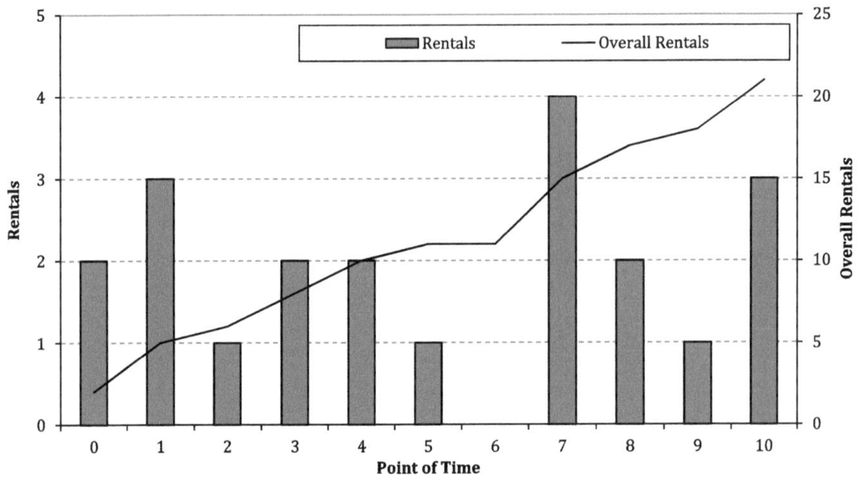

Fig. 4.7 Exemplary Poisson process

in a rental and return demand. Raviv and Kolka (2013) model rentals and returns for a single station according a Poisson distribution. An exemplary Poisson distribution for an expected mean value $\mu = 2$ is depicted in Fig. 4.6. Raviv and Kolka (2013) derive the mean values from historical observations. Considering a set of stations, each trip contains an origin, a duration, and a destination. The sum of the trips results in the rental and return demand for each station and point of time. Brinkmann et al. (2015) model trips by realizations of a Poisson distribution based on expected origin-destination flows derived by Vogel et al. (2011). Trip durations are assumed to be deterministic. The accumulation of Poisson distributed random variables results in a Poisson process for returns and rentals over time. We show the Poisson process exemplarily for rentals in Fig. 4.7. The x-axis shows 11 point of times. On the left y-axis, the rentals per point of time are depicted. These rentals follow the Poisson distribution shown in Fig. 4.6. On the right y-axis, the resulting Poisson process, i.e., the overall rentals over the point of times are shown.

4.5.5 *Requests*

The stochastic models of requests have to consider both customer locations and request times. Usually, locations and times are modeled by two independent random variables. In the following, we first show the spatial modeling, then the temporal modeling of stochastic requests.

Customer locations are modeled twofold. First, they follow a spatial probability distribution in the (continuous) service area. Second, customer requests are modeled by a finite set of possible locations. In urban environments, the number of possible

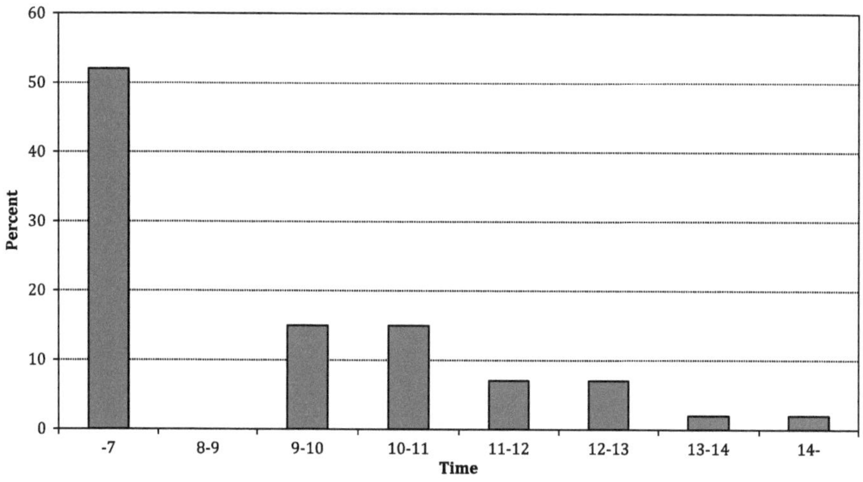

Fig. 4.8 Expected request times (derived from Hvattum et al. 2006)

customer locations is vast. Therefore, possible request locations are not modeled individually, but by a continuous spatial probability distribution. Requests are modeled uniformly distributed (Bertsimas and Van Ryzin 1991) or clustered in certain regions of the service area (Bent and Van Hentenryck 2003). Travel distances and resulting travel times are mainly Euclidean (Bertsimas and Van Ryzin 1991). Ghiani et al. (2009) divide the service area in a set of equidistant zones with equiprobable request possibilities. Bent and Van Hentenryck (2003) and Ulmer et al. (2017b) generate spatial distributions where customers are distributed in clusters. Within each cluster, request probabilities are two-dimensionally normally distributed. (Bent and Van Hentenryck 2004) generate customer regions from Solomon-graphs (Solomon 1987). Every node in the Solomon-graph is transferred to a region. Within the regions, requests follow spatial probability distributions.

Beside spatial probability distributions, customer locations are often modeled by a graph, i.e., a finite set of locations and an according distance matrix. The items of the set may represent known potential customers (Meisel 2011) or representatives of customers in a certain region (Campbell 2006). Because of the vast number of possible customer locations in cities, this representation may in some cases result in a discrepancy between (aggregated) decision and practical implementation and, therefore, of inefficient solutions (Ulmer and Mattfeld 2013). Every customer in the graph is assigned to a request probability. Thomas (2007) assigns every potential customer a uniform probability between 0.1 and 0.75 to request service within the time horizon. The locations and travel times are often derived from the Solomon-instances. Hvattum et al. (2006) use historical observations by Wahl et al. (2002) and Euclidean distances to generate a set of potential customers.

Hvattum et al. (2006) also draw on historical observations for the times of the requests as shown in Fig. 4.8. Usually, a subset of (initial) customers is given in the beginning of routing, defined by the degree of dynamism (Larsen et al. 2002). The times of later requests often result from a Poisson process and are therefore mainly uniformly distributed within the time horizon (Ghiani et al. 2009; Thomas 2007; Ulmer et al. 2015a). Bent and Van Hentenryck (2004) differentiate morning and afternoon requests. Afternoon requests are only allowed in regions close to the depot for feasibility reasons. Therefore, the temporal distribution of requests is heterogeneous.

4.5.6 Stochastic Transitions in SDVRPs

The described stochastic properties have to be transferred into transitions of a Markov decision process. A transition in dynamic vehicle routing generally occurs by the travel of vehicles and the conduction of tasks. The time between two decision points be t_k^{k+1}. During this time, a (sub-)realization of the stochastic properties is given. For an overview of general methods to transfer stochastic information into transitions of the MDP, the interested reader is referred to Powell (2011, p. 63f). In the following, we briefly discuss methods to transfer stochastic information into transitions for SDVRPs.

For stochastic travel times, a transition reveals the actual required travel time until this point of time. For some problems, it also contains determinations of subsequent travel times, e.g., to the next customer. Stochastic service times show the same properties. A transition may reveal the termination of a service, the percentage of service already conducted, or the deterministic time still required to finish the service.

Transitions for demands are twofold. In distribution or collection applications, a transition generally reveals the demand when the vehicle has reached the customers location. In shared mobility, every realized demand in the time span t_k^{k+1} is revealed because the inventories are known. Nevertheless, the destinations of the rented vehicles may be unknown. For stochastic requests, the transition reveals a set of new requests with request times within t_k^{k+1}.

The probabilities of transitions between PDS S_k^x and S_{k+1} can be directly derived by the combination of times of the decision points, PDS' elements and the stochasticity described earlier. Especially for stochastic demands and requests, the modeling as Poisson distribution induces the according Poisson process. This Poisson process can be naturally integrated in the MDP.

4.6 Modeling SDVRPs as MDPs

In this section, we show the process of modeling SDVRPs as MDPs. We define exemplary SDVRPs as Markov decision processes. Notably, models differ regarding the specific SDVRP. In the examples, we focus on SDVRPs without many restrictions to allow a succinct depiction of the models. We first present an SDVRP with stochastic travel times and depict the states, decisions, post-decisions states, and transitions of the according MDP in detail. In comparison to this particular problem, we describe MDPs for three SDVRPs, each containing one of the remaining three drivers of uncertainty. For these problems, we do not depict the entire MDP, but the main differences to the model of the SDVRP with stochastic travel times.

4.6.1 Decision Points

We show the MDP's properties for a specific decision point. Decision point are occur, when exogenous and planning process are synchronized, i.e., when new information is revealed and the plans are adapted. In some cases, decision points directly occur immediately when new information is revealed by the exogenous process. In other cases, decision points occur when a vehicle is free, i.e., it has finished service or delivery at a customer. Then, the driver (or dispatcher) is able to process the new information and has to decide where to go next. The final state S_K is usually reached when all constrains are held (e.g., all customers are served or the time limit is reached) and the vehicles have returned to the depot.

4.6.2 Travel Times

We consider the following SDVRP with stochastic travel times: Vehicles have to serve a set of (known) customers. The vehicles start and end the tour in a single depot. Capacities, time windows, service times, and demands are neglected. During traveling, a change of destination is not possible. The objective is to minimize the overall travel time. Travel times are only revealed when a vehicle arrives at a customer. The travel times to the next customer are then known and deterministic.

In the according MDP, a decision point k occurs, when a vehicle is located at a customer. A state S_k consists of the point of time t_k, depot D, customers $\mathscr{C}$, vehicles $\mathscr{V}$, and the vehicles' statuses $v(k) = (v_1(k), \ldots, v_{|\mathscr{V}|}(k)) \in \{0, 1\}^{|\mathscr{V}|}$ with $v_i(k) = 1$ if vehicle V_i is located at a customer and is therefore free, $v_i(k) = 0$ else.

For every vehicle V_i, the state contains the time of arrival at the destination $t_i^a(k)$. If a vehicle is located at a customer, i.e., $v_i(k) = 1$ and $t_i^a(k) = 0$, a state consists of the vehicles' locations $l_i^v(k) \in \{D, l_j^c : C_j \in \mathscr{C}\}$. If vehicle V_i is currently traveling, i.e., $v_j(k) = 0$ and $t_i^a(k) > 0$, the state contains the next vehicle location $l_j^v(k) \in \{D, l_i^c : C_i \in \mathscr{C}\}$. The random variable for this problem is the travel time matrix $\mathscr{M}^\tau$. In a decision point k, a state consists of a realization $\mathscr{M}_k^\tau$ of this random variable.

Further, a state contains the customers' statuses. The status $r^k \in \{0, 1\}^{|\mathscr{C}|}$ of a customer C_i is $r^k(C_i) = 1$ if the customer still needs to be served and $r^k(C_i) = 0$ if the customer is already served. The set of customers still to serve is $\mathscr{C}_k = \{C \in \mathscr{C} : r^k(C) = 1\}$. Decisions are made about the next customers $C_{\text{next}}^{V_i}(k) \in \mathscr{C}_k$ to visit for vehicles $V_i : v_i(k) = 1$. If $\mathscr{C}_k = \emptyset$, the vehicle's next location is set to

$C^{V_i}_{\text{next}}(k) = D$. The immediate rewards, i.e., costs for this problem are the travel times to the next customers as depicted in Eq. 4.4.

$$R(S_k, x) = \sum_{V_i \in \mathcal{V}: v_i(k)=1} \mathcal{M}^{\tau}_k (l^v_i(k), C^{V_i}_{\text{next}}(k)) \tag{4.4}$$

The elements of the resulting post-decision state S^x_k are similar to S_k. The point of time $t_{k,x}$ of the PDS S^x_k can be depicted twofold. First, $t_{k,x}$ is set to the point of time t_k the decision is made. Second, if applicable, $t_{k,x}$ is set to the point of time the next decision point occurs as depicted in Eq. 4.5.

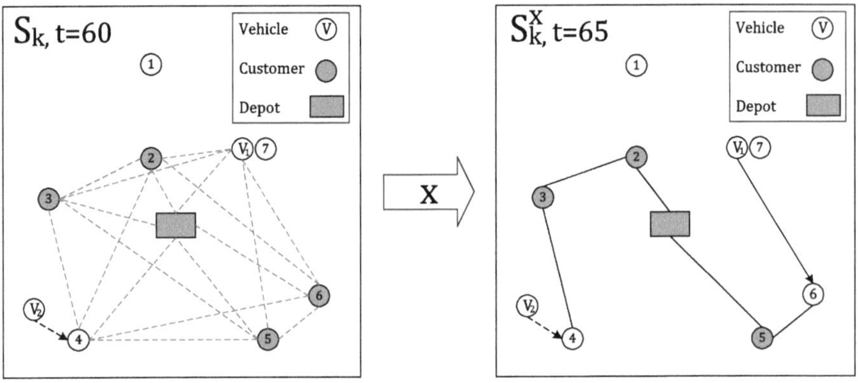

Fig. 4.9 Exemplary MDP-component for an SDVRP with stochastic travel times

$$t_{k,x} = t_k + \min_{V_i \in \mathcal{V}} (v_i(k) \cdot \mathcal{M}^{\tau}_k (l^v_i(k), C^{V_i}_{\text{next}}(k)) + t^a_j(k)) \tag{4.5}$$

The locations of V_i with $v_i(k) = 1$ are set to $l^v_j(k) = C^{V_i}_{\text{next}}(k)$. The arrival times $t^a_i(k, x)$ of V_i are updated by $t^a_i(k, x) = t^a_i(k) - t_{k,x} + t_k$. The status for a vehicle V_i with $t^a_i(k, x) = 0$ is updated accordingly to $v_i(k, x) = 1$. Else, $v_i(k, x)$ is set to zero. The location of V_i is $l^v_i(k, x) = C^{V_i}_{\text{next}}(k)$. The statuses of the next customers are set to zero: $r^{k,x}(C^{V_i}_{\text{next}}) = 0$. The next state S_{k+1} is equal to S^x_k except that a new realization of the travel matrix $\mathcal{M}^k_{C,D}(k + 1)$ is provided. This matrix is unknown before. The MDP terminates if $\mathcal{C}_k = \emptyset$, $l^v_j(k) = D$, and $t^a_i(k, x) = 0 \, \forall \, V_i \in \mathcal{V}_i$.

An exemplary (and simplified) MDP-component in depicted in Fig. 4.9. On the left side, the state S_k is shown. S_k is described by $t_k = 60$, D, $\mathcal{V} = \{V_1, V_2\}$, $\mathcal{C} = \{C_1, \ldots, C_7\}$, $r^k = (0, 1, 1, 0, 1, 1, 0)$, $v(k) = (1, 0)$, $t^a(k) = (0, 5)$, and $l^v(k) = (C_7, C_4)$, $\mathcal{M}^{\tau}_k$. For the purpose of presentation, the known travel time matrix $\mathcal{M}^{\tau}_k$ is not explicitly depicted but indicated by the gray lines. The current point of time is $t_k = 60$ min. Depot D is located in the center of the service area. Two vehicles $\mathcal{V} = \{V_1, V_2\}$ and seven customers $\mathcal{C} = \{C_1, \ldots, C_7\}$ are given. Customers C_1 and C_7 are already served. Customer C_4 will be served by vehicle V_2

in $t_2^a(k) = 5\,\text{min}$. This is indicated by $v_2(k) = 0$ and $l_j^v(k) = C_4$ in state S_k and depicted by the arrow and dashed line in Fig. 4.9. V_1 is currently located at a customer, $l_1^v(k) = C_7, v_1(k) = 1, t_1^a(k) = 0$. A decision x is made about the next customer to be visited for V_1. In this case, $C_6 = C_{\text{next}}^{V_i}(k)$ is selected (with travel time $\mathcal{M}_k^\tau(l_1^v(k), C_{\text{next}}^{V_i}(k)) = 12\,\text{min}$). The immediate costs are $R(S_k, x) = 12$.

The PDS S_k^x is depicted on the right side of Fig. 4.9. S_k^x is defined by $t_{k,x} = 65$, D, $\mathcal{V} = \{V_1, V_2\}$, $\mathcal{C} = \{C_1, \ldots, C_7\}$, $r^{k,x} = (0, 1, 1, 0, 1, 0, 0)$, $v(k, x) = (0, 1)$, $t^a(k, x) = (7, 0)$, and $l^v(k, x) = (C_6, C_4)$. A travel time matrix is not given. The right side of Fig. 4.9 shows a possible plan to finish the routes. For the given problem, this plan is not mandatory because no (time) constraints are given and a feasibility check of decision x is not necessary. The stochastic transition to a next state S_{k+1} contains the travel of the vehicles and a new travel time matrix $\mathcal{M}_{k+1}^\tau$. The following state S_{k+1} is defined by $t_{k+1} = 65$, D, $\mathcal{V} = \{V_1, V_2\}$, $\mathcal{C} = \{C_1, \ldots, C_7\}$, $r^{k+1} = (0, 1, 1, 0, 1, 0, 0)$, $v(k+1) = (0, 1)$, $t^a(k+1) = (7, 0)$, $l^v(k+1) = (C_6, C_4)$, and $\mathcal{M}_{k+1}^\tau$.

4.6.3 Service Times

We consider a SDVRP with stochastic service times. A set of customers is given. The vehicles start at the depot and eventually return to the depot. Capacities, time windows, and demands are neglected. Travel times are deterministic. The objective is to minimize the overall sum of travel and service time. The service time of a customer depends on vehicle and customer and is only revealed when the vehicle finishes service at a customer.

The MDP for stochastic service times is similar to the MDP for stochastic travel times. The main differences are that the service times t^{service} and the time of the next decision point t_{k+1} are unknown and the travel time matrix is deterministic. State S_k and PDS S_k^x additionally contain the point of time a vehicle started the service at a customer. Decisions are made about the next customers to visit. A decision point occurs when a vehicle finishes service at a customer. The rewards (costs) are the sum of deterministic travel time and stochastic service times and therefore themselves random quantities revealed by the transition. The time of the next decision point t_{k+1} is a random variable induced by the individual stochastic service times $t_{i,j}^{\text{service}}$ at a customer C_i where a vehicle V_j is currently located.

4.6.4 Demands

For the considered SDVRP with stochastic demands, vehicles have to serve a set of customers. The vehicles are capacitated and start with a fully loaded fill level at the depot. They have to return to the depot. The fill level at the return can be arbitrary. Time windows and service times are neglected. Each customer has a specific demand. This demand is only revealed when a vehicle arrives at a customer. The demand has to be satisfied by the according vehicle. If the current fill level is not sufficient, the vehicle can satisfy the demand partially, then has to return to the depot for replenishment, and return to the same customer to satisfy the remaining demands. The objective is to satisfy all demands minimizing the overall travel time.

Compared to the MDP of the SDVRP with stochastic travel times, for the SDVRP with stochastic demands, a state S_k and PDS S_k^x additionally consist of the fill levels $c_i^{\text{fill}}(k)$ of the individual vehicles V_i. The fill level $c_{i,j}^{\text{fill}}(k)$ in a state S_k for a vehicle V_i at a customer C_j is the difference between the actual fill level $c_i^{\text{fill}}(k)$ and the (stochastic) customer's demand $c_j^{\text{demand}}(k)$: $c_{i,j}^{\text{fill}}(k) = c_i^{\text{fill}}(k) - c_j^{\text{demand}}(k)$. Therefore, it is modeled as a random variable. If $c_{i,j}^{\text{fill}}(k) < 0$, a depot revisit is mandatory. Otherwise, a depot revisit is optional. Decisions are made about the next customer(s) to visit and a (preemptive) replenishment. The reward (cost) is the resulting travel time of the decision. A post-decision state S_k^x contains the resulting fill levels of the vehicles $c_i^{\text{fill}}(k, x)$.

4.6.5 Requests

For the considered SDVRP with stochastic requests, identical working hours are given for every vehicle. The vehicles start their tours in the depot and have to return within the working hours. Over time, customers request service. Decisions are made about the subset of requests to be confirmed and the next customers to be visited. Not confirmed requests are rejected. The objective is to maximize the number of confirmed requests.

We differentiate two versions of the described problem. In the first version, the set of possible requesting customers is known. In the second version, the requests follow a spatial-temporal probability distribution as described in Sect. 4.5.5. For both versions, a decision point k occurs when a vehicle is located at a customer. Decisions are made about the subset of requests to be confirmed. The immediate reward is the cardinality of the confirmed subset. Both versions demand a feasible plan serving all confirmed customers and returning to the depot within the time limit. This plan is mainly needed for a feasibility check. Dispatchers are not committed to follow the plan in subsequent decision points.

For the first version, a state consists of all possible customers $\mathscr{C}$ and the statuses of a customer C_i are extended as depicted in Eq. 4.6 (Thomas 2007). r^k is a random variable. In a decision state S_k, all cases of statuses are possible. Beside routing, decisions x are made about the subset $(C)_2(k)$ of customers with $r^k(C_i) = 2$, i.e., the customers who requested service between t_{k-1} and t_k. The status of each customer $C_i \in \mathscr{C}_2(k)$ is changed to $r^k(C_i) = 3$ (rejection), $r^k(C_i) = 1$ (confirmation), or $r^k(C_i) = 0$ (confirmation and next customer to serve). The stochastic transition changes a subset of customers C_i with $r^k(C_i) = 4$ to $r^k(C_i) = 2$.

$$r^k(C_i) = \begin{cases} 0 & \text{if } C_i \text{ is already served} \\ 1 & \text{if } C_i \text{ is confirmed but not served} \\ 2 & \text{if } C_i \text{ requested service} \\ 3 & \text{if } C_i \text{ is rejected} \\ 4 & \text{if } C_i \text{ did not request yet} \end{cases} \tag{4.6}$$

For the second version, the possible customers are not known beforehand. The set of requesting customers at decision point k is a random variable. Conferring to Eq. 4.6, status $r^k(C_i) = 4$ is not given. Arguably, even the customers with $r^k(C_i) = 0$ may be neglected in a state definition. A stochastic transition provides a new set of requesting customers $\mathscr{C}_2(k+1)$.

4.7 Vehicle Routing with Recourse Actions

Some vehicle routing problems are modeled as static but include *recourse actions*. For these problems, an a priori plan is determined including possible reactions to revealing information (Errico et al. 2016). The recourse actions are applied in the (dynamic) execution of a plan. Hence, the according model can be described as SDVRP and therefore modeled as MDP. The a priori plan with recourse actions can be seen as a dynamic decision policy as described in Sect. 4.2.2.

4.8 Route-Based Markov Decision Process

Even though MDPs may be a natural model for stochastic and dynamic decision problems, we may experience a disconnect between model and method. More specific, while the routing community is historically conditioned to think in routes as solutions, for MDPs, decisions are usually one made about the next customer to visit. Analyzing the literature revealed that also most of the solution methods determine routes in every decision state. To connect MPDs and SDVRPs, Ulmer et al. (2016a) present a *route-based* MDP formulation. They propose to carry a set of planned routes in the decision states and to augment decisions to route updates. This allows to align routing methods to the MDP terminology. To account for routes in the states and for updates of routes as decisions, it may be helpful to replace the reward function by a *marginal reward*. This marginal reward reflects the difference of conventional rewards between two routing plans. As Ulmer et al. (2016a) show are these two MDP formulations and reward-functions equivalent. Furthermore, the route-based MDP allows a direct connection between model and solution method and may therefore be a valuable tool in the SDVRP-community.

4.9 Implications

In this section, we briefly summarize the properties and advantages of modeling an SDVRP as a MDP. We are positive that the MDP enhances the understanding of an SDVRP regarding the dynamic decision making, the stochastic impact, the reconstruction, definition, and simulation of both SDVRPs and solution methods. MDPs are especially suitable for the derivation of anticipatory methods considering future decision points and stochastic transitions.

4.9.1 Properties of SDVRP

Keeping in mind that dynamic decision making strongly differs from static (e.g., ex post) solutions, the MDP allows the modeling of a problem regarding the perspective of the planner and dispatcher. It explicitly depicts the specific planning situations when a dispatcher has to decide about the next steps. It contains all information the dispatcher has access to in a specific point of time for planning. The MDP highlights the important properties at a specific decision point and provides the immediate impact (and reward) of a decision. In combination with dynamism, stochasticity significantly impacts the decision making. The selection of suitable plans is hereby dependent of the uncertain nature of the problem. The MDP is able to highlight the important global influencing factors of the SDVRP as well as the particular factors for a specific decision point.

4.9.2 Definition, Reconstruction, and Simulation

An SDVRP modeled as an MDP is unambiguously defined by the determination of decision points, states, decision, post-decision states, and transitions. Researchers are able to reconstruct the specific SDVRP. This enables unaffected comparisons of different solution methods for the identical problem. Not only the SDVRP, but also a solution approach defined by the determination of a decision policy can be reconstructed. The modeling as an MDP allows a (faster) understanding of the applied approach, e.g., identifying the point of time the decision procedure is applied or the calculation is conducted.

The modeling as an MDP has also practical advantages. When implementing a simulation framework, programmers can easily follow the MDP in their implementation. As an example, subsequent occurrences of decision points may be implemented by a loop of increasing points of time. Further, states at decision points can be extracted from the simulation and accessed by the solution method. This allows independence of simulated problem and applied decision policy determined by the solution method.

4.9.3 Anticipation and Prescriptive Analytics

We have to be aware that the definition of an SDVRP as an MDP is independent of the applied solution method. Nevertheless, the use of an MDP often facilitates the application of prescriptive analytics to derive anticipatory decision polices. This is enabled by the integration of Bellman's Equation, the access to all state's and PDS's elements in every decision point, and the possibility to include future stochastic transitions. In the next chapter, we draw on the MDP to define and classify anticipation. In Chap. 6, we present anticipatory solution approaches regarding the classification. The solution approaches generally base on the MDP-formulations of the SDVRPs.

Chapter 5
Anticipation

In this chapter, the concept and definition of *anticipation* is recalled and anticipation is embedded in the context of MDP and SDVRP. The main idea of anticipation is to incorporate a predictive lookahead model of future stochasticity and decision making in the solution approach (Butz et al. 2003). Approaches only considering the immediate reward are called *myopic*.

5.1 Definition

The first formal definition of anticipation is given by Rosen in the 1970's (Rosen 2012). Rosen differentiates between *feedback* and *feedforwarding* in decision making problems. Given a decision problem, a feedback is an ex post evaluation of the decision made. So, the decision's quality is solely evaluated after its application. The idea of anticipation is to feedforward. Feedforwarding uses a predictive model of the object system to include realistic decision outcomes in the decision's evaluation before the decision is applied.

A general definition of anticipation in decision problems is given by Schneeweiss (1999). Schneeweiss investigates the interrelations between hierarchical levels of a decision problem. Hierarchies can be used to derive decisions in complex decision problems. Therefore, the decision problem can be decomposed (compare Polya 2014) in two levels, a top-level and a base-level, as shown in Fig. 5.1. Decisions made in the top-level influence the base-level and the combined overall outcome.

Anticipation is defined as the (feedforward) consideration of the base-level's relevant characteristics in the top-level decision. As an example, top-level decision making may consider aggregated base-level information to calculate the base-level's impact on the system outcomes. Beside anticipation, the top-level may provide instructions to the base level to achieve suitable overall outcomes, i.e., rewards.

© Springer International Publishing AG 2017
M.W. Ulmer, *Approximate Dynamic Programming for Dynamic Vehicle Routing*,
Operations Research/Computer Science Interfaces Series 61,
DOI 10.1007/978-3-319-55511-9_5

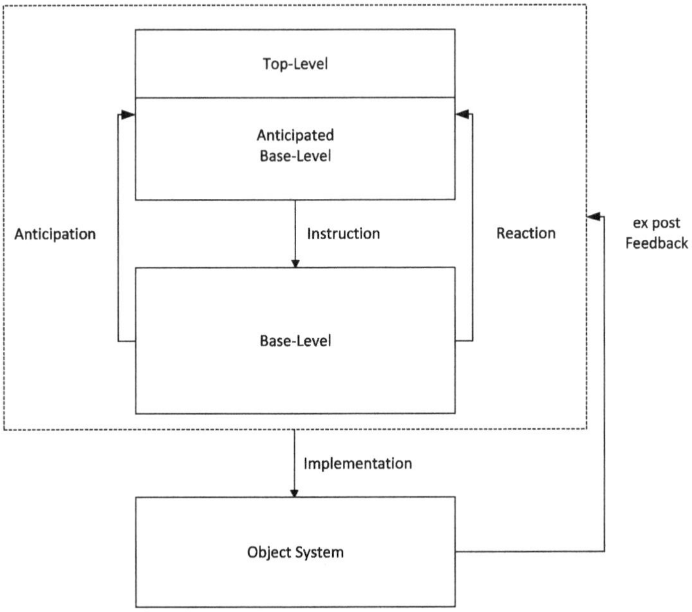

Fig. 5.1 Hierarchical decision problem (Schneeweiss 1999)

Further, as depicted in Fig. 5.1, the base-level may react to the top-level decision by feedback before the actual implementation. The difference between reaction and anticipation is that for reaction a communication between top and base level has to be established. Finally, the decisions are implemented to the object system and the system's outcome provides ex post feedback to the decision maker. Anticipation can be advantageous compared to reaction and feedback, because it does not require actual implementation and communication with the base-level.

5.2 Anticipation in SDVRPs

Schneeweiss (1999) defines anticipation for any decision problem containing different levels of decision making. In the following, we transfer the definition to SDVPRs. An SDVRP can be seen as a special case of a hierarchical decision problem. We describe the special properties of SDVRPs in Fig. 5.2. The figure shows three subsequent decision points for a SDVRP. Top-level and base-level are separated by time. The top-level represents the decision at the current decision point. The base-level represents the future decision point(s). Given a transition, the next decision point becomes top-level, the remaining future decision points base-level. The temporal separation does not allow a feedback or reaction of the base-level before the decision is implemented but only an ex post feedback. Hence, for SDVRPs, anticipation is highly important to achieve effective decisions and high overall outcomes.

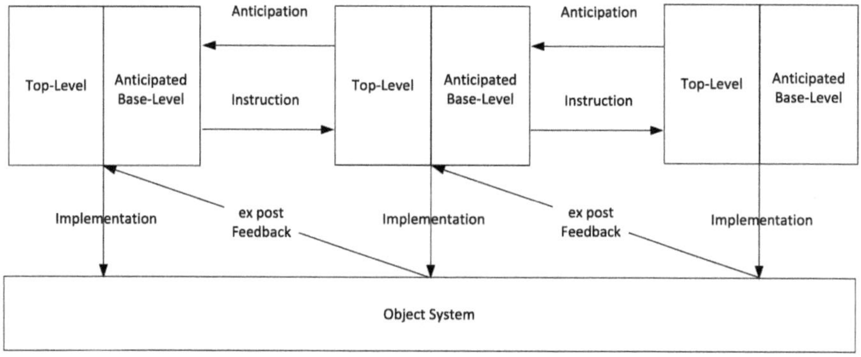

Fig. 5.2 Anticipation in SDVRPs

Based on the definition of Schneeweiss (1999), in the following, we present a classification of anticipation. We start with perfect anticipation. We then classify anticipation in this context. Reactive explicit anticipation allows the explicit consideration of future transitions and decision points to evaluate current decisions and it therefore resembles perfect anticipation. We describe the properties of reactive explicit anticipation in detail in Sect. 5.5.

5.3 Perfect Anticipation

In this section, optimal policies are introduced in Sect. 5.3.1 maximizing the sum of immediate and expected future rewards. To achieve optimal policies, the stochastic dynamic program (SDP, Kall and Wallace 1994) of the MDP can be solved using state values as depicted in Sect. 5.3.2. Nevertheless, solving the SDP is highly limited as described in Sect. 5.3.3.

5.3.1 Optimal Policies

The objective for stochastic dynamic vehicle routing problems is to find an *optimal* decision policy (compare Sect. 4.2.2). In context of Schneeweiss (1999), optimal policies consider every information detail of the base-level within top-level decision making. For deterministic optimization problems, all system information is known at the time of the decision and solutions can be directly evaluated regarding the respective overall rewards. For SDVRPs, the overall reward is only revealed after the application of the decision policy. Further, the reward varies over the problem's realizations depending on the realized stochastic transitions $\omega \in \Omega$. To achieve an optimal policy, decision policies are evaluated regarding the expected future rewards given initial state S_0. The *value* $V^{\pi}(S_0)$ of state S_0 and policy $\pi \in \Pi$ can be calculated as depicted in Eq. 5.1.

$$V^{\pi}(S_0) = \mathbb{E}\left[\sum_{k=0}^{K} R(X_k^{\pi}(S_k))|S_0\right] \tag{5.1}$$

Optimal decision policies $\pi^* \in \Pi$ select decisions leading to the highest expected rewards and therefore maximize the sum of expected rewards as depicted in Eq. 5.2.

$$\pi^* = \underset{\pi \in \Pi}{\arg\max}\ \mathbb{E}\left[\sum_{k=0}^{K} R(X_k^{\pi}(S_k))|S_0\right] \tag{5.2}$$

In a specific state S_k, the optimal decision $X_k^{\pi^*}(S_k)$ can be derived by maximizing the immediate and expected future rewards (*Bellman's Equation*, Bellman 1956) as shown in Eq. 5.3.

$$X_k^{\pi^*}(S_k) = \underset{x \in \mathscr{X}(S_k)}{\arg\max} \left\{ R(x) + \underbrace{\mathbb{E}\left[\sum_{j=k+1}^{K} R(X_j^{\pi^*}(S_j))|S_k\right]}_{=:V^{\pi^*}(S_k,x)} \right\} \tag{5.3}$$

In essence, a policy is called optimal if the policy's values match the maximal expected future rewards.

5.3.2 *Derivation of Optimal Policies*

The maximal expected rewards of a state or a PDS respectively can be used in Eq. 5.3 to achieve optimal policies. Therefore, the expected rewards for every state and PDS have to be calculated. This can be achieved by solving the stochastic dynamic program of the MDP. The values are calculated recursively, starting in K and considering the transition probabilities $\mathbb{P}$ between PDSs and states.

The value of the termination state is $V(S_K) = 0$, because no further rewards follow. The value of PDS S_k^x and state S_k can be calculated using Eqs. 5.4 and 5.5. The calculation of $V(S_k)$ requires the knowledge of all accessible PDSs $S_k^{x_i}$, $\forall x_i \in \mathscr{X}(S_k)$ and the according values $V(S_k^x)$.

$$V(S_k) = \max(V(S_k^x) + R(S_k, x)) \tag{5.4}$$

The value of a PDS can be calculated as seen in Eq. 5.5. Here, the transition probabilities $\mathbb{P}$ and the state values are used.

$$V(S_k^x) = \sum_{S \in \mathscr{S}} \mathbb{P}\left(S_k^x, S\right) \cdot V(S) \quad (5.5)$$

Table 5.1 Anticipation classification

	Non-reactive	Reactive
Implicit	-	$\hat{\mathscr{B}}$
Explicit	$\hat{\mathbb{P}}$	$\hat{\mathscr{B}}, \hat{\mathbb{P}}$
Perfect	n/a	$\mathscr{B}, \mathbb{P}$

Proceeding recursively, starting with the values of termination state S_K, all values can be calculated resulting in an optimal decision policy.

5.3.3 Limitations

The solution of the SDP to achieve optimal policies is computationally tractable only for small-sized MDPs. Given a small set of decision points, possible states, decisions, PDSs, and the according probabilities for transitions, a recursive calculation of the values eventually results in an optimal decision policy. Nevertheless, solving the SDP to SDVRPs of real-world size is highly limited by the number of states and PDSs, decision points, decisions, and transitions (Powell 2011).

The values have to be calculated for every state of the problem. In many real-world settings, the number of states and according decisions is vast or even infinite considering, e.g., numerical state attributes like point of time. The number of decision points defines the number of necessary recursion steps of the SDP. In combination with decisions and possible transitions to consider, the calculation effort increases exponentially. Finally, due to the vast number of states, the probability function $\mathbb{P}$ may not be accessible. As a result, optimal policies cannot be achieved, but only approximated by anticipatory decision policies. In the following, we classify anticipation regarding the degree of anticipation.

5.4 Classification of Anticipation

Schneeweiss (1999) differentiates between reactive and non-reactive anticipation. Further, he differentiates between explicit, implicit, and perfect anticipation. The classification is shown in Table 5.1. $\mathscr{B}$ indicates the application of Bellman's Equation, $\hat{\mathscr{B}}$ indicates the application of a derivation of Bellman's Equation. The use of (some) stochastic information is depicted by $\hat{\mathbb{P}}$. If all stochastic information is considered, the entry contains $\mathbb{P}$.

5.4.1 Reactive Versus Non-reactive

Following Schneeweiss (1999), we define *reactive* anticipation as integration of future developments in the decision making. More specifically, reactive anticipation in context of MDP includes (derivations of) the Bellman's Equation. Reactive approaches estimate the values of states and PDSs. How the estimation is achieved is secondary. Approaches not drawing on Bellman's Equation generally provide *non-reactive* anticipation.

5.4.2 Implicit, Explicit, and Perfect

The approaches are classified by the utilization of stochastic information. Approaches not including stochastic information in decision making are called *implicitly* anticipatory. Approaches including stochastic information are called *explicitly* anticipatory. As a result, perfect anticipation is a special case of explicit and reactive anticipation.

5.4.3 Focus of Anticipation: Offline and Online

For reactive anticipation, we additionally consider the focus of anticipation. We draw on the terminology of Hand et al. (2001, pp. 165ff). We differentiate between anticipating the global "model" of the system behavior and the anticipation for a specific local "pattern", i.e., a particular realized state. The focus determines when the approximation is achieved. We differentiate *offline* and *online* anticipation. In the field of artificial intelligence, the two classes are also called "eager" and "lazy" learning (Hendrickx and Van Den Bosch 2005). An offline learning approach is seen as eager because it approximates values before the occurrence of a specific state. Offline means that (all) states of the system are evaluated a priori within a learning (or tuning) phase with no temporal limitations. This leads to high-level, global anticipation. The values of offline approaches have to be stored to access them in the execution phase of the algorithm. Therefore, the anticipated model is often highly aggregated. Online means that the values are calculated ad hoc for a specific realized state within the execution phase. Online approaches usually have only limited time for calculation but no storage is necessary.

5.5 Reactive Explicit Anticipation

We describe the concept of reactive explicit anticipation in more detail. Therefore, we draw on the definition of perfect anticipation. In a state S_k, an approach providing perfect anticipation solves the SDP to the MDP exactly. For reactive explicit anticipation, the approach solves an auxiliary SDP approximately. The tree of the auxiliary SDP has more than one branch at the first decision node and contains at least one stochastic transition. We explain these constraints in the following.

The first decision node of the tree of the auxiliary SDP represents the decision state S_k in the MDP. Hence, if this node only contains a single branch, the decision to be selected is already determined. A decision making based on anticipation is therefore not given. Second, the tree has to contain at least one branch representing a stochastic transition. This branch enables the integration of stochastic information in the SDP. In cases without this branch, we may achieve reactive implicit anticipation. The procedure of approximately solving the auxiliary SDP is called approximate dynamic programming (ADP). Based on the concept of an auxiliary SDP, we describe the ADP approaches providing reactive explicit anticipation in Sect. 6.2.

Chapter 6
Anticipatory Solution Approaches

For the purpose of presentation, we present approaches of different degrees of anticipation in context of SDVRPs. We utilize the findings of this chapter to allow a succinct literature classification later on. In the following, we briefly describe approaches providing non-reactive anticipation. We present methods for reactive anticipation in detail in Sects. 6.2–6.6.

6.1 Non-reactive Anticipation

In this section, we briefly describe non-reactive anticipatory approaches. First, we present the idea of non-reactive implicit anticipation. Then, we describe non-reactive explicit anticipation. We briefly describe possible shortcomings of non-reactive anticipatory approaches.

6.1.1 Non-reactive Implicit Anticipation

Approaches of non-reactive implicit anticipation neither draw on Bellman's Equation for decision making nor include stochastic information. These approaches mainly exploit assumptions what a good decision policy should look like (compare Powell 2014, policy function approximation). Often, non-reactive implicit approaches implement a rule of thumb and are therefore highly problem-specific. As an example, given an SDVRP with stochastic demands, a non-reactive implicit policy may consider safety buffers for vehicles' fill levels. If the fill level falls beneath this buffer, the depot is revisited for replenishment to ensure efficient fulfillment of future demands. Given stochastic customer requests, waiting strategies are often applied to

© Springer International Publishing AG 2017

M.W. Ulmer, *Approximate Dynamic Programming for Dynamic Vehicle Routing*,
Operations Research/Computer Science Interfaces Series 61,
DOI 10.1007/978-3-319-55511-9_6

allow efficient integration of future customers. Even though stochastic properties are not considered explicitly, non-reactive implicit approaches often provide sufficient decision making with efficient runtimes. Nevertheless, for more complex SDVRPs, the derivation of non-reactive implicit approaches becomes more difficult. General approaches may not be able to consider the problem's and instances' complexity, while tailored solution approaches may result in overfitting.

6.1.2 Non-reactive Explicit Anticipation

These approaches actively incorporate possible future events in current decision making but do not draw on Bellman's Equation. They often include stochasticity by simulation of the future and sampling of realizations. Therefore, they are in many cases computationally expensive. As an example, for stochastic customer requests, waiting strategies may depend on a set of sampled future requests. One prominent approach of non-reactive explicit anticipation is the multiple scenario approach (MSA) by Bent and Van Hentenryck (2004). In every decision point, MSA samples future events (e.g., customers). A set of plans is generated regarding these events. From these plans, a *distinguished plan* is selected based on specific criteria dependent on the problem's characteristics. The derivation of non-reactive explicit approaches is challenging because as non-reactive implicit anticipatory approaches, they require assumptions about "good" decisions or plans and additionally need to include the influence of stochastic information. As a result, they are often highly problem-specific and also bear the risk of overfitting. Further, non-reactive anticipation generally not includes future decision points and possible adaptions of plans and is therefore additionally limited.

6.2 Reactive Anticipation

We are mainly interested in reactive anticipatory approach, since they allow a detailed and problem-specific consideration of future developments. Reactive anticipation is established in many application fields, e.g., in managerial finance. One prominent example is the Black–Scholes model to depict the development of stock prices and to derive optimal prices for derivatives (Black and Scholes 1973). A derivative is a financial product based on a specific stock. The stochastic process of this model is the price development of the underlying stock. States contain of the point of time, the portfolio, and the current stock price. Decisions are made about hedging, i.e., counterbalancing the portfolio to the realized stock prizes by selling or buying stock. The optimal policy minimizes the costs of the hedging and the price of the derivative respectively. The Black–Scholes model changed the stock market and is one foundation for today's algorithmic trading.

Reactive anticipation is achieved by estimating the expected future rewards (or costs) to go for a state. The expected rewards are random variables and depend on both, future decisions and stochastic transitions. They may be derived exactly by means of stochastic programming, for example, by recursive calculation. Still, problems oriented on real world applications are usually highly complex in many dimensions of the MDP, e.g., in the number of possible states, decisions, or stochastic transitions. These complexities are known as the "Curses of Dimensionality". As a result, an exact calculation of the expected rewards is not possible. Heuristic methods of reactive anticipation aim on *approximating* the expected rewards for a specific state. This can be achieved by means of approximate dynamic programming (ADP, Powell 2011). The field of ADP-methods is vast. In essence, methods of ADP draw on (forward) programming, often via simulation, to estimate the expected rewards for a state. ADP has been successfully applied to a vast number of real-world applications. Most recently, methods of ADP are utilized to anticipate future developments in the energy market (Powell and Meisel 2016).

Reactive anticipatory approaches can be differentiated by how and when the estimation is conducted. In the following, we briefly present the idea of reactive implicit approaches in Sect. 6.2.1. In Sect. 6.2.2, we describe reactive explicit approaches. Therefore, we recall the concept of approximate dynamic programming. In Sect. 6.3, we use ADP to describe the concept of lookaheads approximating values ad hoc within the execution phase. We present the rollout algorithm (RA) as online lookahead approach in detail. We focus on post-decision RAs and show methods to reduce the required calculation effort.

In Sects. 6.4 and 6.5, we present offline approaches of value function approximation. We focus on methods of approximate value iteration (AVI) and show the generally required steps of state space aggregation and partitioning to achieve a sufficient approximation of the states' values. We introduce new concepts of state space partitioning to allow an efficient and effective approximation. In Sect. 6.6, we define a hybrid method combining RA and AVI to exploit the advantages of both offline and online approximation.

6.2.1 *Reactive Implicit Anticipation*

Reactive implicit approaches do not draw on future events for the estimation of the value, but use states' attributes and assumptions about the problem to calculate the states' values (compare Powell 2014, cost function approximation). No simulation or sampling of realizations is conducted, but the values are calculated by a predefined set of weighted functions regarding the attributes. In SDVRPs with stochastic demands, future rewards (here, costs) can be derived by the gap between fill level and safety buffer. If the fill level is only slightly higher than the safety buffer, the expected costs are generally relatively high compared to a large gap. Reactive implicit approaches allow a consideration of states' and decisions' properties. Nevertheless, attributes, functions, and parameters have to be determined without considering the actual problem's behavior. Therefore, complex dependencies within the problem and instances can often not be considered sufficiently. Further, the functions are often induced by assumptions about the problem's behavior which may not be accurate in the actual implementation.

6.2.2 Reactive Explicit Anticipation

Reactive explicit approaches evaluate the problem's (post-decision) states. They solve the according SDP not perfectly, but by approximation. The main challenges for reactive approaches are the *curses of dimensionality* (Powell 2011). Mainly, reactive explicit anticipation is achieved by the application of approximate dynamic programming described in Sect. 6.2.3. For a successful explicit reactive anticipation, the original SDP has to be reduced by maintaining the program's main characteristics. Even though the reduction has to be problem specific, there are several generic methods, presented in Sect. 6.2.4.

6.2.3 Approximate Dynamic Programming

The expected future rewards are accumulated to the value of a post-decision state and depend on the applied policy. The value for any given policy π can be recursively calculated applying decision rule X_j^π in all subsequent decision points $j = k+1, \ldots, K-1$ for all possible transitions given a post-decision state S_k^x. The achieved value is called the (system) value $V_s^\pi(S_k^x)$ of the regarding post-decision state depending on π as depicted in Eq. 6.1.

$$V_s^\pi(S_k^x) = \mathbb{E}\left[\sum_{j=k+1}^{K} R(S_j, X_j^\pi(S_j)) | S_k^x\right] \tag{6.1}$$

As shown in Eq. 5.3, the decision rule of an optimal policy π^* maximizes the sum of the immediate reward and the value of the post-decision state. Equation 5.3 can be reformulated as shown in Eq. 6.2.

$$X_k^{\pi^*}(S_k) := \underset{x \in X(S_k)}{\arg\max} \left\{ R(S_k, x) + V_s^{\pi^*}(S_k^x) \right\} \tag{6.2}$$

Conversely, the assignment of specific values to all post-decision states defines a decision policy if Eq. 6.2 is applied. ADP aims on finding a policy π^a close to an optimal policy. Instead of directly approximating a policy π^a to the optimal policy π^*, methods of ADP aim on approximating the expected rewards $V_s^{\pi^*}$ by the values V^{π^a}. This can be achieved by approximately solving the according SDP. Therefore, the original SDP is reduced and aggregated. In the following, we briefly discuss methods for reduction and aggregation of the SDP.

6.2.4 Reducing the SDP

In the following, reduction and aggregation techniques and their impacts to state space, decision space, transition space, and decision points are presented. For an extensive overview of ADP, the interested reader is referred to Powell (2011). ADP approaches usually draw on a subset of the methods.

State Space

A state or PDS contains all achievable problem information. For many SDVRPs, the amount of information is vast, resulting in a (nearly) infinite number of possible states and post-decision states respectively. To apply ADP, instead of considering every state by itself, states can be aggregated and partitioned to calculate identical values for similar states. *Aggregation* is mainly achieved by neglecting state information or by differentiating states regarding a set of key attributes. *Partitionings* group aggregated states to a single value. As an example, instead of continuous points of time, time-intervals are used. Further, the state space can be reduced by exclusion of states (e.g., via sampling). Especially, areas of the state space are excluded which might be unlikely to be observed or overall unpromising, i.e., which might contain states with low values.

Decision Space

Decisions often contain several *partial decisions*. In many SDVRPs, decisions are made about customers assignments and the vehicles' sequences of customers. Every additional partial decision increases the decision space exponentially. To apply ADP, the decision space can be reduced using *decomposition* (Polya 2014). Decomposition divides a problem in a top-level and a base-level sub-problem. While the top-level sub-problem is considered in detail, the base-level sub-problem are solved subsequently. As a result, only the top-level partial decisions have to be evaluated while the base-level partial decisions might follow a heuristic. Further, decisions can be aggregated. In these cases, decision making is not represented on the highest level of detail. As an example, instead of routing decisions containing assignments and sequences, only an assignment problem is solved using problem-relaxation (Spivey and Powell 2004). The according solution is then transferred to the overall routing decision by applying a subsequent sequencing algorithm.

Transition Space

In many cases, the number of possible transitions is vast or even infinite. As an example, customers may request service at arbitrary locations in the service area. To reduce the number of transitions, single transitions are sampled representing the overall set of possible transitions (Bertsekas and Tsitsiklis 1996). Sampling additionally reduces the number of considered states significantly. Further, the transitions can be aggregated. For example, instead of every possible realization of a continuous demand, significant demand levels ("low", "medium", "high") can be used.

Decision Points

The number of decision points to consider defines the number of recursion steps to calculate. Hence, every additional decision point multiplies the SDP to solve. To apply ADP, decision points can be combined to an aggregated set of decision points. So, for example, instead of minute by minute decision making, only hourly decision points are considered. Further, the number of decision points can be reduced by neglecting decision making and rewards after a certain point of time in the SDP. This would cut the entire SDP after this point of time.

6.2.5 Resulting Approaches

The combination of aggregation and reduction methods leads to a set of possible ADP-approaches. The intuitive idea is to solve the SDP approximately online at a specific decision state. Therefore, we briefly present the concept of lookaheads in Sect. 6.3. We present one lookahead-approach, the rollout algorithm, in detail. RAs maintain the state's full level of detail. RA draws on reduction of state space and transition space by sampling and reduces the decision space by applying a base policy within the SDP.

In Sects. 6.4 and 6.5, we present the two most prominent offline ADP-methods for post-decision state evaluation: approximately modeling the value function (MVF) and approximate value iteration. Both methods use simulation runs within an (offline) learning phase to approximate the values. The approximated values then induce a policy which can be applied in the (online) execution phase. We present AVI in detail. AVI generally maintains the decision space's dimensionality but draws on aggregation and partitioning within the state space and reduction of the transition space via sampling and simulation. We give an elaborate overview of the strategies for partitioning of the state space. This is required to describe our proposed method of a dynamic lookup table adapting to the approximation process. In Sect. 6.6, we propose a hybrid RA combining online RA and offline AVI. We show for an exemplary problem the advantages of hybrid RAs compared to solely offline and online ADP-methods.

6.3 Lookahead and Rollout Algorithm

In this section, we briefly recall the concept of lookaheads (Powell 2011, p. 225f). The idea of lookahead-methods is to approximately solve the SDP in a specific state for a limited number of decision points and cut the SDP afterwards. If applicable, the resulting SDP is solved exactly. In cases, where the decision space is small, but the number of realizations and transitions is vast, sampling is applied. For SDVRPs, usually both decision and transition space are vast. Hence, the SDP has to be additionally reduced regarding the number of considered decisions. This is mainly conducted by applying rollout algorithms (Bertsekas et al. 1997). Lookaheads are online and can be seen as “local” anticipatory approaches providing detailed anticipation for a specific state but are often not able to consider the overall SDP (Hendrickx and Van Den Bosch 2005).

The rollout algorithm provides online reactive explicit anticipation. For an extensive definition and classification of RAs, the interested reader is referred to Goodson et al. (2014). An RA can be seen as a derivation of the *Pilot* method for dynamic decision problems (Duin and Voß 1999). The main feature of an RA is the application of a *base policy* to reduce the SDP.

RAs usually consider a limited time of the future in a high level of detail. The idea of RA is to solve a highly limited SDP online to achieve a *tendency* of the expected rewards for different decisions. κ-step RAs maintain the high level of detail of the states for κ decision points, but reduce the number of possible decisions and transitions afterwards. After decision point κ, decisions in the SDP are selected by a base policy. For most of the SDVRPs, not every possible realization can be considered. Sampling and simulation are applied (Bertsekas and Tsitsiklis 1996). Instead of solving the SDP combined with the base policy exactly, for every (post-decision) state, n transitions are sampled. For a κ-step RA, this sampling has to be done κ times iteratively, i.e., within each decision point in the considered SDP. This may consume significant amounts of computation time. The resulting average rewards are used for the PDS-evaluation. RAs allow an explicit state consideration without aggregation, but on the expense of high online computation times and a significant reduction in decision and transition space. One advantage of RAs is that the approach does not require an exact evaluation of states but only an order of states regarding the values. Notably, RAs can be seen as a method to integrate reactive anticipation in a given policy. For a specific class of base policies, so called “sequentially consistent” policies, an improvement of the solution quality by the RA is guaranteed. Ulmer et al. (2017a) proved that the RA provides at least as good solution quality as the base policy. Further, the RA enables reactive anticipation even for myopic decision policies and is therefore especially suited for SDVRPs.

In the following, we define the RA for post-decision state evaluation in Sect. 6.3.1. Because of the usually vast number of possible realizations and transitions in SDVRPs, we focus on post-decision RAs. Here, sampling and the base policy are applied instantly for every PDS. The solution quality of RAs is highly dependent on the number of sample runs. Because the RA is applied in the execution phase, the available calculation time is limited. We propose the use of the *fully sequential procedure for indifference-zone selection* (IZS) by Kim and Nelson (2001) to reduce the number of required sample runs and calculation time in Sect. 6.3.2.

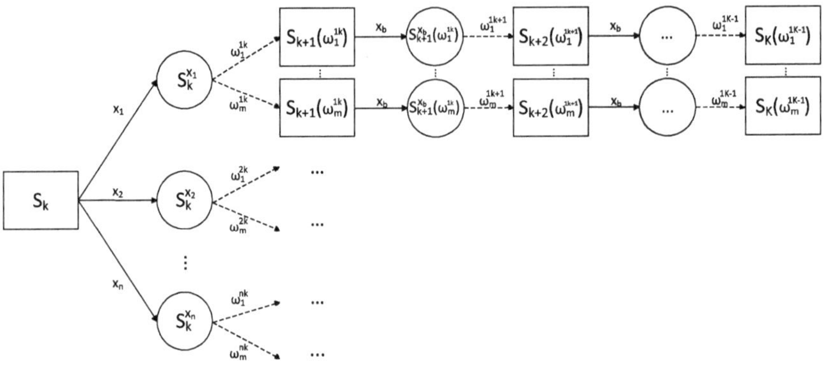

Fig. 6.1 Post-decision rollout algorithm

6.3.1 Functionality

Figure 6.1 shows the SDP of an RA for post-decision state evaluation (Ulmer et al. 2015b). The RA contains a base policy π_b. Further, a number of sample runs m has to be determined. Given a state S_k, a set of possible decisions $x_1, \ldots, x_n$, and the resulting set of PDSs $S_k^{x_1}, \ldots, S_k^{x_n}$, let $(\omega_1, \ldots, \omega_m) = \bar{\Omega} \subset \Omega$ be a set of sampled realizations. As we can see, the resulting decision tree consists of $n \times m$ independent straightforward branches. Each branch begins in the sampled state $S_{k+1}(\omega_i)$ of the first transition ω_i^k induced by realization ω_i and terminates in the final state $S_K(\omega_i^{K-1})$. Decision making is conducted by the base policy π_b. The resulting decisions are $X^{\pi_b}(S_{k+j})$ indicated by x_b in Fig. 6.1. Transitions are generated by the realization ω_i.

Algorithm 1 describes the procedure of decision making for a post-decision RA. Given a state S_k, decisions $x_1, \ldots, x_n$, the according rewards $R(S_k, x_j)$, $j = 1, \ldots, n$, the PDSs $\mathscr{P}_k = (S_k^{x_1}, \ldots, S_k^{x_n})$, a set of sampled realizations $\{\omega_1, \ldots, \omega_m\}$, and the base policy π_b, then, for every PDS S_k^x, the RA simulates the remaining decision points of every realization ω_i. In the decision points, the decision induced by π_b is applied. The overall reward of a PDS $\hat{V}(S_k^x)$ is the average of the rewards per realization. The RA selects the decision x^* leading to the maximum sum of immediate reward $R(S_k, x^*)$ and expected future rewards $\hat{V}(S_k^{x^*})$.

6.3.2 Efficient Computing: Indifference Zone Selection

RAs are online approaches. The evaluation is conducted in the execution phase of the algorithm. In many cases, the provided time for calculation is highly limited. Even though parallelization may reduce the overall required runtime (Guerriero and Mancini 2003), efficient RAs make use the provided time for a more effective evaluation. Approaches to reduce runtime are the *optimal computing budget allocation*

Algorithm 1: Post-Decision Rollout Algorithm

Input: S_k, $\{x_1, \ldots, x_n\}$, $R(S_k, x_j)$, $j = 1, \ldots, n$, $\mathscr{P}_k = (S_k^{x_1}, \ldots, S_k^{x_n})$, $\{\omega_1, \ldots, \omega_m\}$, π_b
Output: Decision x^*

for all $S^x \in \mathscr{P}_k$ **do**
  $i \leftarrow 0$
  $\hat{V}(S^x) \leftarrow 0$
  `// Simulation`
  while $(i \leq m)$ **do**
    $S \leftarrow S_k$
    $S^x \leftarrow S_k^x$
    while $(S^x \neq S_K)$ **do**
      $S \leftarrow (S^x, \omega_i)$
      $x \leftarrow X^{\pi_b}(S)$
      $\hat{V}(S_k^x) \leftarrow \hat{V}(S_k^x) + \frac{1}{m} R(S, x)$
      $S^x \leftarrow (S, x)$
    end
    $i \leftarrow i + 1$
  end
  `// Selection`
  $R^* \leftarrow 0$
  for all $S_k^x \in \mathscr{P}_k$ **do**
    if $(R(S_k, x) + \hat{V}(S_k^x) \geq R^*)$ **then** $x^* \leftarrow x$;
  end
end
return x^*

(Powell 2011), distributing computational time to parallelized sample runs and the *fully sequential procedure for indifference-zone selection* by Kim and Nelson (2001) terminating sample runs for inferior states. IZS is easily implemented and has already proven to allow significant runtime reductions for dynamic vehicle routing problems (Ghiani et al. 2009). In the following, we describe the idea and functionality of IZS in detail.

IZS can significantly reduce the required calculation time of RAs. Kim and Nelson (2001) have analyzed the theoretical functionality of IZS assuming normally distributed value realizations. Even though the values may follow a different distribution, the functionality of IZS can be transferred. For example, IZS for routing was successfully applied by Ghiani et al. (2009) for a short-term sampling strategy. IZS can be applied for any RA. In the following, we describe the functionality for post-decision states because we apply IZS in a post-decision RA in Sect. 10.8.

In a nutshell, the idea of IZS is to subsequently simulate realizations and terminate the evaluation of *dominated* post-decision states. States are dominated if their estimated reward-to-go is significantly lower in comparison to other states. The allowed difference declines over the number of simulated realizations. The free calculation time resulting from the termination can then be utilized to sample more frequent realizations for promising post-decision states.

The original procedure by Kim and Nelson (2001) requires three parameters: n_0, α, and δ. $n_0 \geq 2$ defines the number of sample runs for an initial evaluation. $1-\alpha$ defines the confidence interval of the probability that the selected post-decision state has the largest sum of reward and value. δ defines the *indifference zone*, i.e., the neglectable value difference. Usually, small α and δ increase the required number of sample runs and according runtime. The impact of n_0 on the required sample runs is not easily depictable. A small n_0 may not allow the termination of any post-decision state evaluation and result in many additional calculations and comparisons. A large n_0 requires a large number of initial sample runs for all states but may avoid the further evaluation of inferior post-decision states afterwards. The original IZS is applied until only one candidate remains. RAs do not require the distinction of a single *best* post-decision state value, but only a sufficient approximation for every post-decision state value. To allow termination with several remaining states, we consider a maximal number of sample runs $n_{\max}$ in the procedure as alternatively proposed by Kim and Nelson (2001).

IZS is described in Algorithm 2. Given a set of post-decision states $\mathscr{P}_k = \{S^{x_1}, \ldots, S^{x_m}\}$ to evaluate, let $R(S, x_j)$ be the reward of x_j. Let further α and δ be given with $0 \leq \alpha \leq 1$ and $0 \leq \delta \in \mathbb{R}_+$. Let $\hat{V}^i_{\text{run}}(S^{x_j})$ be the realized value for S^{x_j} in sample run i. After sample run i, let $\hat{V}^i(S^{x_j})$ be the average value of S^{x_j} as depicted in Eq. 6.3.

$$\hat{V}^i(S^{x_j}) = \frac{1}{i}\sum_{r=1}^{i} \hat{V}^i_{\text{run}}(S^{x_j}) \tag{6.3}$$

Let $\mu^i(S^{x_j})$ be the sum of reward and value, and for $i \geq 2$, let $\sigma_i^2(S^{x_j}, S^{x_k})$ be the observed variance of the difference between S^{x_j}, S^{x_k} as depicted in Eq. 6.4. Let $\mathscr{S}^i$ be the set of remaining post-decision states after sample run i.

$$\sigma_i^2(S^{x_j}, S^{x_k}) = \frac{\sum_{r=1}^{i}\left(\left[\hat{V}^i_{\text{run}}(S^{x_j}) - \hat{V}^i_{\text{run}}(S^{x_k})\right] - \left[\hat{V}^i(S^{x_j}) - \hat{V}^i(S^{x_k})\right]\right)^2}{i-1} \tag{6.4}$$

$\sigma_i^2(S^{x_j}, S^{x_k})$ indicates the correlation between two post-decision state values. In the extreme case given correlation of 1 for S^{x_j} and S^{x_k}, $\sigma_i^2(S^{x_j}, S^{x_k})$ is zero. There are several approaches of IZS with different update procedures for σ_i^2 (and h_i^2). In the proposed algorithm, we update the parameters after each sample run as proposed by Kim and Nelson (2006).

IZS contains three phases: *Initialization*, *Screening*, and *Termination*. In the initialization phase, n_0 sample runs for every post-decision state are conducted. This

Algorithm 2: Indifference Zone Selection

Input: $S, \{x_1, \ldots, x_n\}, R(S, x_j), \mathscr{P}_k = (S^{x_1}, \ldots, S^{x_n}), n_0, n_{\max}, \alpha, \delta$

Output: Approximate Values $\hat{V} = \{\hat{V}(S^{x_j}), \forall j = 1, \ldots, n\}$

// Initialization
Run $n_0 - 1$ Sample Runs for all $S^x \in \mathscr{P}_k$
$i \leftarrow n_0 - 1$
$\mathscr{P}^i \leftarrow \mathscr{P}_k$
// Screening
while $(i \leq n_{max}) \vee (|\mathscr{P}^i| \geq 2)$ **do**
  $i \leftarrow i + 1$
  for all $S^x \in \mathscr{P}^i$ **do**
    Run Sample Run $\hat{V}^i_{\text{run}}(S^x)$
    $\hat{V}^i(S^x) \leftarrow \frac{1}{i} \sum_{r=1}^{i} \hat{V}^i_{\text{run}}(S^{x_j})$
    $\mu^i(S^x) \leftarrow R(S, x) + \hat{V}^i(S^x)$
  end
  $h_i^2 \leftarrow \left[\left(\frac{2\alpha}{|\mathscr{P}^i|-1}\right)^{-2/(i-1)} - 1\right](i-1)$
  for all $S^{x_j}, S^{x_k} \in \mathscr{P}^i$ **do**
    $\sigma_i^2(S^{x_j}, S^{x_k}) \leftarrow (i-1)^{-1} \sum_{r=1}^{i} \left(\left[\hat{V}^i_{\text{run}}(S^{x_j}) - \hat{V}^i_{\text{run}}(S^{x_k})\right] - \left[\hat{V}^i(S^{x_j}) - \hat{V}^i(S^{x_k})\right]\right)^2$
    $W^i_{jk} \leftarrow \max\left\{0, \frac{\delta}{2i}\left(\frac{h_i^2 \sigma_i^2(S^{x_j}, S^{x_k})}{\delta^2} - i\right)\right\}$
  end
  $\mathscr{P}^{i+1} \leftarrow \emptyset$
  for all $S^{x_j} \in \mathscr{P}^i$ **do**
    $a \leftarrow 0$
    for all $S^{x_k} \in \mathscr{P}^i$ **do**
      if $\mu^i(S^{x_j}) < \mu^i(S^{x_k}) - W^i_{jk}$ **then**
        $\hat{V}(S^{x_j}) \leftarrow \hat{V}^i(S^{x_j})$
        $a \leftarrow 1$
        break
      end
    end
    if $a = 0$ **then** $\mathscr{P}^{i+1} \leftarrow \mathscr{P}^{i+1} \cup \{S^{x_j}\}$
  end
end
// Termination
for all $S^x \in \mathscr{P}^{n_{max}}$ **do**
  $\hat{V}(S^{x_j}) \leftarrow \hat{V}^{n_{\max}}(S^{x_j})$
end
return $\hat{V}$

Table 6.1 State properties and sample runs

States	S^{x_1}	S^{x_2}	S^{x_3}	S^{x_4}	
V	1.0	2.0	4.0	6.0	
R	3.0	1.0	3.0	2.0	
$V+R$	4.0	3.0	7.0	8.0	
σ^2	0.0	1.0	3.0	4.0	
Runs	$\hat{V}^i_{\text{run}}(S^{x_1})$	$\hat{V}^i_{\text{run}}(S^{x_2})$	$\hat{V}^i_{\text{run}}(S^{x_3})$	$\hat{V}^i_{\text{run}}(S^{x_4})$	h_i^2
$i=1$	1.0	3.5	5.8	4.7	–
$i=2$	1.0	2.6	2.7	2.2	55.3
$i=3$	1.0	0.6	5.0	7.8	13.0
$i=4$	1.0	1.8	4.3	1.5	8.5
$i=5$	1.0	3.0	2.5	3.6	7.0
$i=6$	–	–	4.9	10.6	2.2
$i=7$	–	–	4.3	7.5	2.1
$i=8$	–	–	1.9	9.2	2.1
$i=9$	–	–	3.9	5.7	2.1
$i=10$	–	–	–	–	–

results in average values $V^{n_0}(S^{x_j})$ and a sample variance $\sigma^2_{n_0}(S^{x_j}, S^{x_k})$ of the value difference for all tuples of post-decision states.

In the screening phase, states are examined for possible domination. Parameters h_i^2 indicates the reliability of the already achieved μ^i and σ_i^2. h_i^2 is dependent on the confidence interval α, the number of post-decision states still to consider, and the number of sample runs already conducted. An increase in sample runs leads to a decrease in h^2. Parameters W^i_{jk} span the corridor, in which μ^i of two states S^{x_j} and S^{x_k} are allowed to differ. If a state violates the lower bound of the corridor, it is dominated by the compared state and terminated, i.e., excluded from the evaluation. W^i_{jk} depends on the reliability parameter h_i^2, the sample variance $\sigma_i^2(S^{x_j}, S^{x_k})$, the indifference zone δ, and the number of sample runs i. The smaller h_i^2 and $\sigma_i^2(S^{x_j}, S^{x_k})$ and the larger i are, the smaller is W^i_{jk} resulting in a narrower corridor. A small h_i^2 indicates a high reliability of the values. A small $\sigma_i^2(S^{x_j}, S^{x_k})$ indicates a similar behavior of the two states. This allows for a more confident determination of a possible domination. Dominated states are terminated.

After each iteration of the screening phase, IZS checks if a termination criterion is satisfied. This is either the case if only one post-decision state remains or if the maximal number $n_{\max}$ of sample runs is reached. In the first case, the remaining post-decision state induces the selected decision. In the second case, the decision leading to the post-decision state with the largest $\mu^{n_{\max}}$ is selected.

Table 6.2 IZS process

	μ	σ^2				W_{jk}				$\mu^j - \mu^k + W_{jk}$			
		S^{x_1}	S^{x_2}	S^{x_3}	S^{x_4}	S^{x_1}	S^{x_2}	S^{x_3}	S^{x_4}	S^{x_1}	S^{x_2}	S^{x_3}	S^{x_4}
$i = 5$													
S^{x_1}	4.0	–	1.3	2.1	6.1	–	0.4	0.9	3.8	–	1.1	−2.1	1.8
S^{x_2}	3.3	1.3	–	3.9	10	0.4	–	2.2	6.5	−0.3	–	−1.5	3.8
S^{x_3}	7.1	2.1	3.9	–	4.6	0.9	2.2	–	2.7	4.0	6.0	–	3.8
S^{x_4}	6.0	6.1	10.0	4.6	–	3.8	6.5	2.7	–	5.7	9.1	1.6	–
$i = 6$													
S^{x_3}	7.2	1.8	3.5	–	9.3	0.0	0.2	–	1.2	3	4.2	–	1.3
S^{x_4}	7.1	12.3	17.0	9.3	–	1.8	2.6	1.2	–	4.8	6.5	1.1	–
$i = 7$													
S^{x_3}	7.2	1.5	2.9	–	8.5	0.0	0.0	–	0.8	2.9	4.0	–	0.6
S^{x_4}	7.4	11.1	14.9	8.5	–	1.2	1.8	0.8	–	4.6	6.0	1.0	–
$i = 8$													
S^{x_3}	6.9	1.9	3.2	–	11.9	0.0	0.0	–	1.1	2.7	3.6	–	0.1
S^{x_4}	7.9	11.3	14.5	11.9	–	1.0	1.4	1.1	–	4.9	6.1	2.0	–
$i = 9$													
S^{x_3}	6.9	1.7	3.0	–	10.4	0.0	0.0	–	0.7	2.6	3.7	–	0.2
S^{x_4}	7.9	9.9	12.8	10.4	–	0.6	1.0	0.7	–	4.5	5.7	1.6	–

Example: Indifference Zone Selection

In the following, we provide a short example illustrating the functionality of IZS. Let us consider a single decision point, a given state, and four decisions x_j, $j = 1, \ldots, 4$. The quantities of the resulting post-decision states $\mathcal{P} = \{S^{x_j}, j = 1, \ldots, 4\}$ are depicted in the upper part of Table 6.1. The value in every PDS S^{x_j} follows a normal distribution with expected value $V(S^{x_j})$ and variance σ^2. $V(S^{x_1})$ is therefore deterministic. Additionally, we provide the immediate reward R for all four decisions. We assume the value realizations of S^{x_3} and S^{x_4} are correlated by 0.2 and the rest of the realizations to be independent.

The parameters are set as $\alpha = 0.2$, $\delta = 1.0$, $n_{\min} = 5$, and $n_{\max} = 10$. In the initialization phase, five sample runs are conducted. For $i > n_0$, only sample runs for the remaining $S^{x_j} \in \mathcal{P}^i$ are required. All (required) sample runs and the according h_i^2 are depicted in the lower part of Table 6.1. As we can see, sampling of S^{x_1}, S^{x_2} stops after 5 sample runs and the overall process terminates after 9 sample runs. In the following, we analyze the individual sample runs regarding μ, σ^2, W_{jk} and the resulting IZS-steps. The process is depicted in Table 6.2.

At first, $n_0 = i = 5$ sample runs are conducted, indicated in the first segment of Table 6.2. Parameter μ is calculated as the sum of the immediate rewards and the average of the five realizations. In the first five sample runs, h_i^2 declines. For $i = 5$, h_5^2 is still large because of the small number of sample runs and the large cardinality of $\mathscr{P}^5$. This results in relatively large W-corridors. Nevertheless, S^{x_1} is dominated by S^{x_3}. Further, S^{x_2} is dominated by S^{x_1} and S^{x_3}. As a result, these two states are eliminated leading to a new set $\mathscr{P}^6 = \{S^{x_3}, S^{x_4}\}$.

Between $i = 6$ and $i = 8$, h_i^2 drops significantly because the number of considered states is small and the sample runs increase. As a result, the W-corridors tighten. In $i = 9$, S^{x_3} is dominated by S^{x_4} and is eliminated. Only S^{x_4} remains and $|\mathscr{P}^9| = 1$ results in termination.

The example allows us to identify two statements. First, IZS can identify the lack of a sufficient number of sample runs. S^{x_3} provides the largest $\mu^5(S^{x_3}) = 7.1$ after five sample runs, while the theoretically most valuable state S^{x_4} provides only $\mu^5(S^{x_4}) = 6.0$. A decision according to μ^5 after 5 runs would therefore provide not the best expected solution quality. Therefore, more sample runs have to be conducted. Second, the number of required sample runs is reduced by 30%. Half of the states are eliminated right after n_0 sample runs. IZS is able to identify the optimal decision and the whole process terminates before $n_{\max}$ is reached.

Rollout algorithms allow a comprehensive consideration of the particular state to be evaluated. This allowed a detailed evaluation for every observed state in the state space. Since the number of states in SDVRPs are manifold, RAs are in particular suitable for SDVPRs. Still, RAs have to be applied online during the execution phase. The calculation time available is highly limited. IZS allows an early rejection of inferior PDS-candidates and therefore enables an effective use of the limited calculation time.

6.4 Value Function Approximation

In many cases, online calculation time for approximation is highly limited. VFA-approaches approximate the value for a state or PDS offline. After the offline approximation (or learning) phase, the achieved values can be efficiently accessed in the online execution phase of the algorithms. VFAs are offline and can be seen as "global" anticipatory approaches providing a general anticipation of the overall SDP. Often, VFAs are not able to consider local anticipation for a specific state (Hendrickx and Van Den Bosch 2005).

We differentiate two types of VFAs. The first one aims on approximately modeling the value function based on regression analysis (MVF). The second one approximates the values for individual states via approximate value iteration. MVF is usually efficient, but in many cases highly limited due to the regression analysis model. In the following, we briefly describe MVF. For an extensive overview on MVF, the interested reader is referred to (Powell 2011, p. 235ff). AVI allows a more independent evaluation but often on the expense of high calculation effort. We present AVI in detail in Sect. 6.5.

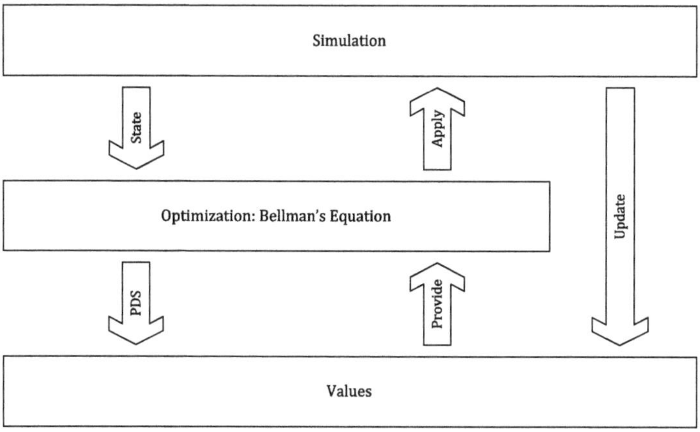

Fig. 6.2 Approximate value iteration (Soeffker et al. 2016)

Approximately modeling the value function can be seen as *parametric* VFA. The idea of MVF is to approximate the real value function by using *basis functions*. These basis functions draw on a set of problem-specific key attributes to calculate expected future rewards. As an example, the expected future rewards highly depend on the state's point of time. The attribute point of time interrelates with the expected future rewards. The exact relation is, however, unknown. So, the value $\tilde{V}$ is the weighted combination of different basis functions ϕ_f, $f \in \mathcal{F}$ based on the attributes as shown in Eq. 6.5. The weights θ_f are approximated over the simulation runs, e.g., by multiple linear regression (Freedman 2009).

$$\tilde{V}(S_k^x) = \sum_{f \in \mathcal{F}} \theta_f \phi_f(S_k^x) \tag{6.5}$$

The main advantage of MVF is that only the manageable number of weights has to be approximated. As a result, efficient approximation can be achieved by only a few simulation runs. Nevertheless, basis functions have a main disadvantage. They have to be defined beforehand and cannot be changed during the approximation process. So, they contain a priori assumptions about the system's behavior, i.e., the interrelations between attributes and expected rewards. These assumptions may limit the quality of the achieved policies significantly. For many SDVRPs, the values of states depend on a variety of interrelated influencing factors like customer sequence or vehicle's load. MVF may not be able to reveal the resulting complex dependencies and the detailed structure of the value function (Bertsimas and Demir 2002).

6.5 Approximate Value Iteration

In contrast to MVF, approximate value iteration approximates independent and individual values for post-decision states. This can be seen as *non-parametric* VFA.

Algorithm 3: Approximate Value Iteration

Input: Initial Values V^{π_0}, Realizations $\bar{\Omega} = \{\omega^1, \ldots, \omega^m\}$, Step Size α
Output: Values V^{π_m}

// Initialization
$i \leftarrow 1$
while $(i \leq m)$ **do**
  $k \leftarrow 0$
  $S_0 \leftarrow S_0^{\omega^i}$
  $x \leftarrow \underset{x \in X(S_0)}{\arg\max} \left\{ R(S_0, x) + V^{\pi_{i-1}}(S_0^x) \right\}$
  $S_0^x \leftarrow (S_0, x)$
  $R_k \leftarrow R(S_0, x)$
  $\mathscr{P} \leftarrow \{S_0^x\}$
  // Simulation
  while $(S_k^x \neq S_K)$ **do**
    $k \leftarrow k + 1$
    $S_k \leftarrow (S_{k-1}^x, \omega_{k-1}^i)$
    $x \leftarrow \underset{x \in X(S_k)}{\arg\max} \left\{ R(S_k, x) + V^{\pi_{i-1}}(S_k^x) \right\}$
    $S_k^x \leftarrow (S_k, x)$
    $R_k \leftarrow R_{k-1} + R(S_k, x)$
    $\mathscr{P} \leftarrow \mathscr{P} \cup \{S_k^x\}$
  end
  // Update
  $R_{-1} \leftarrow 0$
  for all $S_k^x \in \mathscr{P}$ **do**
    $V^{\pi_i}(S_k^x) \leftarrow (1 - \alpha) V^{\pi_{i-1}}(S_k^x) + \alpha (R_K - R_k)$
  end
  $i \leftarrow i + 1$
end
return V^{π_m}

The process and the components of AVI are shown in Fig. 6.2. For a more general overview of AVI-methods, the interested reader is referred to (Powell 2011, p. 127ff). As MVF, AVI uses a simulation for approximation. At decision points, the state and the corresponding decisions are extracted and evaluated regarding the current values and Bellman's Equation. The resulting decision is applied in the simulation. After a simulation run, the values are updated according to the simulation's realized values. Subsequently, the individual PDSs' values are approximated to the real values.

The procedure in detail is described in Algorithm 3. AVI starts with initial values V^{π_0} and a resulting policy π_0. Then, AVI simulates the problem by subsequently sampling a subset of the problem realizations $\bar{\Omega} \subset \Omega$. Within each simulation run i, for the given realization $\omega^i \in \bar{\Omega}$, policy π_{i-1} induced by the values $V^{\pi_{i-1}}$ is applied (*exploitation*). In some cases, decisions may be selected randomly to force an *exploration* of the state and decision space. After each simulation run i, the values are updated with the realized rewards $R_{k+1}, \ldots, R_{K-1}$ after PDS S_k^x according to Eq. 6.6. Parameter α defines the step size of the approximation process.

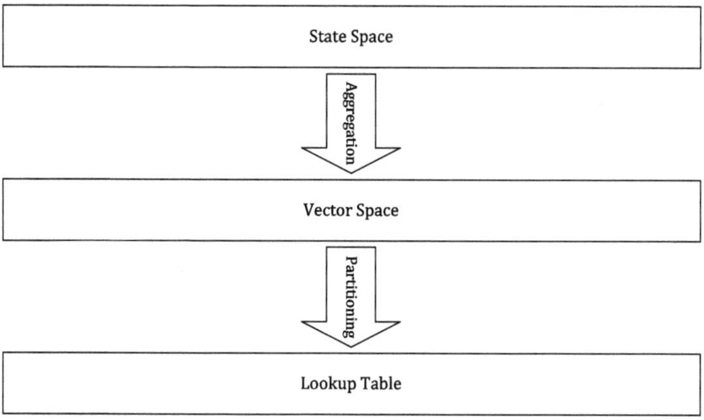

Fig. 6.3 Aggregation and partitioning of a state space

$$V^{\pi_i}(S_k^x) = (1-\alpha)V^{\pi_{i-1}}(S_k^x) + \alpha \sum_{i=k+1}^{K-1} R_i \tag{6.6}$$

The final policy $\pi_{|\bar{\Omega}|}$ is then applied in the execution phase.

6.5.1 Post-decision State Space Representation

Every value calculated by AVI in the learning phase has to be stored to access it in the execution phase. For some problems, an individual evaluation of every post-decision state may be possible. Generally, the number of post-decision states $S_k^x \in \mathcal{P}$ is large. For AVI, the more frequent a value is accessed and updated, the more accurate is its approximation. If values are only accessed sparsely, the approximation and the solution quality of the achieved policy might be impaired (Barto 1998, p. 193). The number of value accesses depends on the number of observations of the particular states. As a result, instead of applying AVI directly on the states, AVI is applied on a (post-decision) *state space representation*. The state space representation has to allow efficient storage, effective approximation, and a resulting high-quality decision making. Figure 6.3 shows the required steps of aggregation and partitioning of the state space to achieve a *lookup table* as sufficient state space representation.

In the following, we describe the idea of aggregation to assign groups of post-decision states to vectors. A vector contains key attributes of the post-decision states. Then, we present methods to partition the achieved vector space to a lookup table for evaluation. If the key attributes are numerical, a canonical vector space is given. Otherwise, a metric on the vector set is needed to create a vector space and for partitioning. This means that a distance can be calculated for every vector-pair. For an overview on metric spaces, the interested reader is referred to Kelley (1975, p. 118ff).

For the purpose of presentation, we assume the key attributes to be numerical and the vector space to be Euclidean. Nevertheless, the presented partitioning strategies are applicable for every metric space.

6.5.2 Aggregation

To achieve a sufficient approximation, some dimensions of the high-dimensional post-decision state space are neglected or represented by a set of key attributes v_i, $i = 1, \dots, \nu$ (compare Sect. 6.4). Generally, an *aggregation* $\mathfrak{A} : \mathscr{P} \rightarrow \mathbb{N}^{\nu-p} \times \mathbb{R}^p$ is used to represent post-decision states $\mathfrak{A}(S_k^x) = p^k$ by ν-dimensional vectors $p^k = (v_1, \dots, v_\nu) \in \mathfrak{A}(\mathscr{P}) = \mathscr{Q}$. $\nu - p$ of the dimensions are ordinal or of integer value, p are continuous. The resulting representation $\mathscr{Q}$ can be seen as a ν-dimensional vector space.

In the decision making, the value of a post-decision state can now be represented by the value of the vector p^k: $V_s^\pi(S_k^x) \approx V^\pi(\mathfrak{A}(S_k^x)) = V_q^\pi(p^k)$, with $V_q^\pi(p^k)$ the value of vector p^k. The use of $\mathfrak{A}$ results in a representation $\mathscr{Q}$ significantly smaller than $\mathscr{P}$.

6.5.3 Partitioning: Lookup Table

If all attributes of $\mathscr{Q}$ are of integer value, $\mathscr{Q}$ can be imagined as a ν-dimensional lookup table. In each entry of this table, the value of the according vector p^k is stored. During the learning phase, the LT-entries are updated. The entry values of a LT induce a decision policy regarding Eq. 6.2.

In many cases, even the number of vectors is vast or even infinite assuming continuous key attributes. Since AVI requires a frequent observation of the lookup table entries, the large cardinality of representation $\mathscr{Q}$ may impede the approximation process (Barto 1998, p. 193). To achieve a sufficient approximation, a *partitioning* $\mathscr{I}$ of the vector space is necessary. A partitioning $\mathscr{I}$ groups vectors to entries of a (partitioned) lookup table $\mathscr{E}$. Every entry is then evaluated by approximate value iteration. A partitioned LT $\mathscr{E}$ is defined by a partitioning $\mathscr{I} : \mathscr{Q} \rightarrow \mathscr{E}$ grouping vectors $p_1, \dots, p_q \in \mathscr{Q}$ to a set represented by a LT-entry $\eta = \mathscr{I}(p_1) = \dots = \mathscr{I}(p_q)$. A partitioning is usually defined a priori and is static. We describe different partitioning strategies in Sects. 6.5.5–6.5.7.

We experience a tradeoff between coarse and fine-grained partitionings. A coarse partitioning may result in a fast but insufficient approximation because heterogeneous states are evaluated equally. A fine-grained partitioning may result in sparse observations of states and therefore an inefficient approximation process. We show the tradeoff between efficient approximation and effective approximation in a short example in Sect. 6.5.4. To combine the advantages of coarse and fine-grained partitionings,

Ulmer et al. (2017b) propose the dynamic lookup table described in Sect. 6.5.7. The DLT is a partitioning approach dynamically adapting the partitioning to the approximation process and the problem specifics.

6.5.4 *Efficient Approximation Versus Effective Decision Making*

In this section, we show the impact of partitioning for the single-stage example introduced in Sect. 6.3.2. To show the impact of the number of entries on the approximation process, we specifically analyze the required number of simulation runs for a sufficient approximation. We consider a detailed partitioning $\mathscr{I}$ resulting in LT $\mathscr{E}$ with four entries $\mathscr{E} = \{p_1, \ldots, p_4\}$. Additionally, we consider a partitioning $\bar{\mathscr{I}}$ resulting in a LT with only a single entry $\bar{\mathscr{E}} = \{\bar{p}\}$. In $\bar{\mathscr{E}}$, $\bar{p}$ represents $p_1, \ldots, p_4$. The value in every entry in $\mathscr{E}$ follows a normal distribution with expected value $V(p_i)$ and standard deviation $\sigma^2(p_i)$. Additionally, let ψ_i be the probability of observing p_i. The according values are shown in Table 6.3. The entry values behave heterogeneously, the expected values and deviations rise from p_1 to p_4. The according values of $\bar{p}$ are a result of the partitioning.

Table 6.3 Expected entry values and deviation

Entry	p_1	p_2	p_3	p_4	$\bar{p}$
V	1.0	2.0	4.0	6.0	3.0
σ^2	0.0	1.0	3.0	4.0	4.3
ψ	0.2	0.3	0.4	0.1	1.0
n_i^*	0	1,537	4,610	6,146	6,590

To show the impact of the number of entries on the approximation process, we calculate the expected necessary number of observations n_i^* for every entry and the total number of simulation runs n^* for termination, i.e., for a sufficient approximation in every entry. As a termination criterion, we allow a difference of the average values $\hat{V}$ to the expected values up to 0.05. Further, we calculate the number of required observations $\bar{n}$ for entry $\bar{p}$ and compare the results. For each entry, we calculate the distribution of the expected realizations' average (i.e., $\alpha = \frac{1}{n}$). Then, we derive the probability $\mathbb{P}_k$ that the average lies in the allowed deviation range after k entry observations. We determine n_i^* as the minimal number of observations with probability higher than $\mathbb{P}_k > 95\%$. Let $\hat{V}_l(p_i)$ be the value of the lth entry realization of p_i. Then, the minimum number of observations for entry p_i can be calculated as shown in Eq. 6.7.

$$n_i^* = \arg\min_{k \in \mathbb{N}^+} \left\{ \mathbb{P}_k \left(\left| V(p_i) - \frac{1}{k} \sum_{l=1}^{k} \hat{V}_l(p_i) \right| \leq 0.05 \right) \geq 0.95 \right\} \quad (6.7)$$

The total number of required simulation runs n^* is the maximum number of individual observations considering the probability of observing the entry as depicted in Eq. 6.8.

$$n^* = \max_{i \in \{1,\ldots,4\}} \left\{ \frac{n_i^*}{\psi_i} \right\} \quad (6.8)$$

Sparsely visited entries increase the necessary number of simulation runs of the algorithm. In the example, p_4 requires the largest number of observations for termination with $n_4^* = 6{,}146$. Due to the probability $\psi_4 = 0.1$ of observing p_4, the expected number of runs for termination of the whole process is $n^* = \frac{n_4^*}{0.1} = 61{,}460$. For $\mathcal{I}$, a sufficient approximation is expected after 61,460 simulation runs.

We now show that partitioning $\bar{\mathcal{I}}$ can reduce the number of required simulation runs significantly. The expected value of $\bar{p} \in \bar{\mathcal{E}}$ is the weighted sum of the single expected values as depicted in Eq. 6.9.

$$V(\bar{p}) = \sum_{i=1}^{m} \psi_i V(p_i) = 3.0 \quad (6.9)$$

The variance $\sigma^2(\bar{p})$ can be calculated as shown in Eq. 6.10.

$$\sigma^2(\bar{p}) = \mathbb{E}V(\bar{p})^2 - (\mathbb{E}V(\bar{p}))^2 = \sum_{i=1}^{4} \psi_i V(p_i)^2 - \left(\sum_{i=1}^{4} \psi_i V(p_i) \right)^2 = 4.3 \quad (6.10)$$

The probability distribution of $V(\bar{p})$ is the weighted sum of the single distributions. The number of necessary observations to achieve a maximal deviation of 0.05 with a probability of at least 95% is $\bar{n} = 6, 590$. The number of necessary simulation runs is reduced by 89.3% compared to $\mathcal{E}$. For our example, partitioning $\bar{\mathcal{I}}$ allows a significantly faster approximation. Nevertheless, the partitioning has a large impact on decision making. As we can see from Eq. 6.10, the variance of $V(\bar{p})$ exceeds the variance of all original entries $p_1, \ldots, p_4$. Partitioning $\bar{\mathcal{I}}$ results in a rise of the deviation of the entry value $\bar{p} \in \bar{\mathcal{E}}$. We additionally experience a bias $|V(\bar{p}) - V(p_i)|$ of up to 3.0 for all former entries. Using $\bar{\mathcal{I}}$ results in a less accurate representation and may lead to ineffective decisions.

Evidently, partitioning $\bar{\mathcal{I}}$ results in a suboptimal solution quality. We consider a decision point S and two possible decisions x_a, x_b. Decision x_a leads to entry p_1, x_b to p_4. The immediate rewards are $R(S, x_a) = 2.0$ and $R(S, x_b) = 1.0$. Considering Bellman's Equation, given $\mathcal{E}$, the overall values are $R(S, x_a) + V(p_1) = 3.0$ and $R(S, x_b) + V(p_4) = 7.0$. Hence, we choose x_b and achieve an expected overall outcome of 7.0. With LT $\bar{\mathcal{E}}$, the two decisions result in the same entry $\bar{p}$. Hence, decision x_a is chosen with an overall outcome of 3.0. Due to partitioning $\bar{\mathcal{I}}$, we experience a loss of 4.0. In essence of the example, $\bar{\mathscr{I}}$ allows a faster approximation, but simultaneously leads to a loss in solution quality. We experience a tradeoff between accuracy and approximation efficiency.

6.5.5 Equidistant Lookup Table

Usually, partitionings generate LTs with sets of equidistant intervals. For example, the point of time is represented by intervals of 5 min. The resulting LT of partitioning $\mathscr{I}_l$ for an interval length of l is $\mathscr{E}_l$. An exemplary LT-entry $\eta \in \mathscr{E}_l$ consists of a set of vectors as depicted in Eq. 6.11.

$$\eta = \{(v_1, \ldots, v_\nu) : \rho_i \leq v_i < \rho_i + l, \forall i = 1, \ldots, \nu\} \tag{6.11}$$

For entry η, ρ_i is the lower bound of key attribute v_i. $\rho_i + l$ is the upper bound of v_i.

If we assume discrete key attributes in $\mathscr{Q}$, i.e., $\mathscr{Q} = \mathscr{E}_1$, a partitioning with equidistant interval length l substantially reduces the number of entries as depicted in Eq. 6.12.

$$|\mathscr{E}_l| = l^{-\nu} * |\mathscr{Q}| \tag{6.12}$$

The resulting LT $\mathscr{E}_l$ is significantly smaller than $\mathscr{Q}$ and may allow a faster approximation. Nevertheless, an a priori definition of the intervals may impede the solution quality because heterogeneous states are assigned to the same value as shown in the example earlier. A suitable interval length may differ given different instances and even within an instance. For example, the required level of time-consideration may vary during the day. As a result, every equidistant interval selection provides a suboptimal approximation. In the following, we present methods which allow heterogeneous consideration of state space areas.

6.5.6 Weighted Lookup Table

Equidistant and static partitionings may impede the approximation process. Frequently observed areas of the vector space are not considered in enough detail, while sparsely observed areas are too separated for reliable approximation. To combine the advantages of different interval lengths $l_1, \ldots, l_L$ and to allow differing levels of detail, George et al. (2008) and Powell (2009) suggest to combine multiple partitionings $\mathscr{I}_1, \ldots, \mathscr{I}_L$ and the resulting LTs $\mathscr{E}_1, \ldots, \mathscr{E}_L$. Let $V^{l_i}(\mathscr{I}_i(p))$ be the value of $\eta_{l_i} = \mathscr{I}_i(p)$ in LT $\mathscr{E}_{l_i}$ of a weighted LT (WLT) $\mathscr{E}_{\mathrm{WLT}} = \bigcup_{i=1}^{L} \mathscr{E}_{l_i}$. The overall value $V_{\mathrm{WLT}}(p)$ is calculated as the weighted sum of the single LT-values with weights

$w_{l_1}, \ldots, w_{l_L}$ for every LT as depicted in Eq. 6.13.

$$V_{\mathrm{WLT}}(p) = \sum_{i=1}^{L} w_{l_i}(\mathscr{I}_i(p)) V^{l_i}(\mathscr{I}_i(p)) \tag{6.13}$$

The weights are calculated considering the experienced variance within the entries $\sigma_{l_i}^2(\mathscr{I}_i(p))$, the number of observations $N_{l_i}(\mathscr{I}_i(p))$, and the bias $\mu_{l_i}(\mathscr{I}_i(p))$ of entry $\mathscr{I}_i(p)$. The bias $\mu_{l_i}(\mathscr{I}_i(p)) = |V^{l_i}(\mathscr{I}_i(p)) - V^{l_1}(\mathscr{I}_1(p))|$ is defined by the difference between value $V^{l_i}(\mathscr{I}_i(p))$ and the value of the lowest partitioning level $V^{l_1}(\mathscr{I}_1(p))$. A formula for the weights calculating the total variation of the error and considering all three impacts is provided by Powell (2009) as depicted in Eq. 6.14.

$$w_{l_i}(\mathscr{I}_i(p)) \propto \left(\frac{\sigma_{l_i}^2(\mathscr{I}_i(p))}{N_{l_i}(\mathscr{I}_i(p))} + \mu_{l_i}(\mathscr{I}_i(p))^2 \right)^{-1} \tag{6.14}$$

The weighting favors LTs with large intervals in the beginning for a fast first approximation. Later, LTs with small intervals are weighted higher to achieve a more differentiated evaluation. The weight w_{l_i} for a LT $\mathscr{E}_i$ is increased by a relatively small variance and bias and a large number of observations. In the beginning, the frequently visited entries in the LTs with large intervals allow a fast first estimation of the value function. During the subsequent approximation process, the weights of the more detailed LTs with small intervals increase because of the relatively small variance and bias. Further, WLT allows avoiding ineffective LT-areas. For instance, areas in the LT providing relatively low expected future rewards are early excluded from the approximation process. Hence, the approximation is focused on the effective areas. WLT may allow a faster approximation in the beginning and high quality solutions in the end without any tuning necessary. Nevertheless, WLT leads to increased memory consumption due to the large number of entries. Instead of a single LT, L LTs of different partitioning levels are required.

6.5.7 Dynamic Lookup Table

To exploit the advantages of the WLT and combine those with an efficient approximation, Ulmer et al. (2017b) introduced a third concept: the dynamic LT (DLT). The main idea of DLT is to analyze the problem's behavior and to adapt the approximation process accordingly (compare Meisel and Mattfeld 2010). DLT adapts to the problem specifics regarding value variance and number of entry observations and only requires a single LT. Compared to WLT, this significantly reduces the memory consumption. To the best of our knowledge, a method of ADP dynamically partitioning a lookup table with numerical parameters regarding the entry observations and value deviation has not been presented in the literature. Nevertheless, the idea

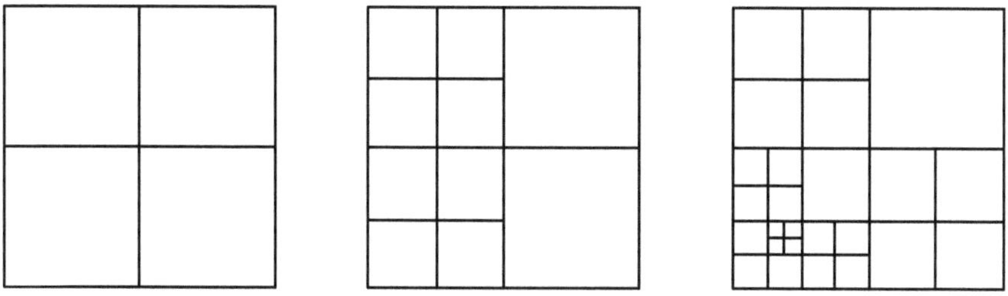

Fig. 6.4 Example for the dynamic evolution of a partitioning $\mathscr{I}$ over the approximation process for $\nu = 2$

of dynamic state space partitioning is not new. A similar concept of dynamically changing the partitioning has been introduced by Bertsekas and Castañon (1989). For an infinite-horizon problem with $|\mathscr{S}| = 100$ and a known transition probability matrix $\mathbb{P}$, Bertsekas and Castañon (1989) partition states regarding their residuals and achieve faster approximation processes. The presented approach is motivated by Bertsekas and Castañon (1989). Since state spaces of SDVRPs are usually vast and the transition functions in many cases not accessible, a direct transfer of this approach is not possible.

DLT adapts the partitioning $\mathscr{I}$ according to the approximation process. DLT $\mathscr{E}^0$ starts with partitioning $\mathscr{I}^0$ with large intervals and subsequently decreases the interval lengths in some areas during the approximation process. In the beginning, a fast, coarse-grained approximation of the few initial entries is provided. During the approximation, the interval length is decreased for some "interesting" areas, i.e., an entry is separated in a set of new entries resulting in a dynamic partitioning $\mathscr{I}^j$ and LT $\mathscr{E}^j$ in simulation run j. Hence, the DLT allows a fine-grained approximation within these areas. For areas of no interest, the partitioning stays in the original design. An evolution of a two-dimensional DLT is exemplified in Fig. 6.4 (Ulmer et al. 2017b). Notably, the DLT for only two dimensions can be associated with a quadtree (Finkel and Bentley 1974), a tree structure for the efficient storage of data. The concept of quadtrees was already combined with Q-learning by Clausen and Wechsler (2000). They propose to use different levels of detail for fractal image compression. In the left, initial partitioning, the intervals are homogenous. During the simulation runs, some areas are considered in more detail, as seen in the central partitioning. In the final partitioning, seen on the right, the lower left area is highly separated and considered in detail, while the large interval lengths of the initial partitioning in the upper right corner remain. Beside the advantage of a fast and effective approximation process, DLT may also allow deriving insights of the problem and solution structure. Areas with a high separation may indicate important areas or general structures for the problem.

Entry Selection

An entry $\eta = \mathscr{I}^j(p)$ of partitioning $\mathscr{I}^j$ and DLT $\mathscr{E}^j$ can only be considered in more detail if a sufficient number of entry observations is given. Otherwise, the new entries may not be observed frequently enough to derive a reliable approximation. Therefore, we consider the number of observations $N(\eta)$ to decide whether we are able to consider η in more detail. Further, entries which are frequently observed have an essential impact on the entire approximation process. The evaluation of those entries has to be very accurate and a detailed consideration is desirable. A more detailed consideration of an entry is additionally required if states of heterogeneous value are grouped in a single entry. This is indicated by a high deviation within the entry's value, represented by the standard deviation $\sigma(\eta)$.

To achieve a method independent of problem and instances, we use the relative values $\frac{N(\eta)}{\bar{N}}$ and $\frac{\sigma(\eta)}{\bar{\sigma}}$. $\bar{N}$ and $\bar{\sigma}$ are the averages of all entries $\eta_i \in \mathscr{E}^j$. A relatively large number of observations is indicated by $\frac{N(\eta)}{\bar{N}} > 1$, a relatively large standard deviation by $\frac{\sigma(\eta)}{\bar{\sigma}} > 1$. A separation of an entry η is conducted if Eq. 6.15 is satisfied.

$$\frac{N(\eta)\,\sigma(\eta)}{\bar{N}\,\bar{\sigma}} \geq \zeta \tag{6.15}$$

A multiplication of the two components allows deterministic entries without deviation to remain unseparated. Further, entries with only a few observations are not split allowing a reliable approximation. The tuning parameter ζ defines the separation frequency of the DLT. A small ζ results in a fast separation of many entries, a large ζ only selects entries with outstanding characteristics in both number of observations and deviation. For a comparison of the impact of N and σ on the approximation process, we define three DLT-approaches: DLT(N, σ) considering both attributes and DLT(N), DLT(σ) only considering a single attribute in Eq. 6.15.

Functionality

The procedure of AVI combined with DLT is depicted in Algorithm 4. Differences compared to plain AVI are aggregation $\mathfrak{A}$, partitioning $\mathscr{I}$, and the resulting LT $\mathscr{E}$. Instead of directly approximating the value of a PDS S^x, the values of the resulting entries η are approximated. Further, the partitioning and LT are updated after every simulation run regarding σ and N. The calculation of σ and the update of $\mathscr{I}$ and $\mathscr{E}$ are described in the following.

Update of Sigma

The standard deviation σ of a set of n observations with realized values $V_1, \ldots, V_n$ and average value $\bar{V}$ is usually calculated as depicted in Eq. 6.16.

Algorithm 4: AVI and Dynamic Lookup Table

Input: Aggregation $\mathfrak{A}$, Initial Partitioning $\mathscr{I}^0$, DLT $\mathscr{E}^0$, Initial Values V^{π_0}, Realizations $\{\omega^1, \ldots, \omega^m\}$, Step Size α, Threshold ζ
Output: Values V^{π_m}, Partitioning $\mathscr{I}^m$, DLT $\mathscr{E}^m$

`// Initialization`
$i \leftarrow 1, \bar{N} \leftarrow 0, \bar{\sigma} \leftarrow 0$
for all $\eta_k \in \mathscr{E}^0$ **do**
  $\sigma(\eta_k) \leftarrow 0, N(\eta_k) \leftarrow 0$
end
`// Simulation`
while $(i \leq m)$ **do**
  $k \leftarrow 0$
  $S_0 \leftarrow S_0^{\omega^i}$
  for all $x \in X(S_0)$ **do**
    $\eta_0^x \leftarrow \mathscr{I}^0\left(\mathfrak{A}(S_0^x)\right)$
  end
  $x \leftarrow \arg\max_{x \in X(S_0)} \left\{R(S_0, x) + V^{\pi_{i-1}}(\eta_0^x)\right\}$
  $S_0^x \leftarrow (S_0, x)$, $R_k \leftarrow R(S_0, x)$
  $\mathscr{P} \leftarrow \{\mathscr{I}^0\left(\mathfrak{A}(S_0^x)\right)\}$
  while $(S_k^x \neq S_K)$ **do**
    $k \leftarrow k + 1$
    $S_k \leftarrow (S_{k-1}^x, \omega_{k-1}^i)$
    for all $x \in X(S_k)$ **do**
      $\eta_k^x \leftarrow \mathscr{I}^{i-1}\left(\mathfrak{A}(S_k^x)\right)$
    end
    $x \leftarrow \arg\max_{x \in X(S_k)} \left\{R(S_k, x) + V^{\pi_{i-1}}(\eta_k^x)\right\}$
    $S_k^x \leftarrow (S_k, x)$
    $R_k \leftarrow R_{k-1} + R(S_k, x)$
    $\mathscr{P} \leftarrow \mathscr{P} \cup \left\{\mathscr{I}^{i-1}\left(\mathfrak{A}(S_k^x)\right)\right\}$
  end
  `// Update`
  $R_{-1} \leftarrow 0$
  for all $\eta_k^x \in \mathscr{P}$ **do**
    $V^{\pi_i}(\eta_k^x) \leftarrow (1 - \alpha)V^{\pi_{i-1}}(\eta_k^x) + \alpha(R_K - R_k)$
    $N(\eta_k^x) \leftarrow N(\eta_k^x) + 1, \sigma(\eta_k^x) \leftarrow \mathit{UpdateSigma}(\sigma(\eta_k^x), R_K - R_k)$
    $\bar{N} \leftarrow \bar{N} + 1, \bar{\sigma} \leftarrow \mathit{UpdateSigma}(\bar{\sigma}, R_K - R_k)$
  end
  for all $\eta_k^x \in \mathscr{P}$ **do**
    if $\frac{N(\eta_k^x)\sigma(\eta_k^x)}{\bar{N}\bar{\sigma}} \geq \zeta$ **then** $\mathit{UpdatePartition}(\mathscr{I}^{i-1}, \eta_k^x)$
    $\mathit{UpdateTable}(\mathscr{E}^{i-1}, \eta_k^x)$
  end
  $i \leftarrow i + 1$
end
return $V^{\pi_m}, \mathscr{I}^m, \mathscr{E}^m$

$$\sigma = \sqrt{\frac{1}{n}\sum_{i=1}^{n}(V_i - \bar{V})^2} \tag{6.16}$$

The standard deviation σ has to be recalculated in every simulation run. A recalculation regarding Eq. 6.16 would require the storage of every realized value V_i. In many cases, the number of simulation runs of AVI is large and an efficient storage not possible. The *Steiner Translation Theorem* allows a reformulation of Eq. 6.16 as depicted in Eq. 6.17.

$$\sigma = \sqrt{\sum_{i=1}^{n} V_i^2 - \frac{1}{n}(\sum_{i=1}^{n} V_i)^2} \tag{6.17}$$

As a result, only the values of $\sum_{i=1}^{n} V_i^2$ and $\sum_{i=1}^{n} V_i$ have to be stored. These values can be efficiently updated given a new observation.

Update of Partitioning and Table

In the update process of $\mathscr{I}^j$, entries $\eta = \mathscr{I}^j(p)$ satisfying Eq. 6.15 are separated to a set of 2^ν new entries $\eta_1 = \mathscr{I}_1^{j+1}(p), \ldots, \eta_{2^\nu} = \mathscr{I}_{2^\nu}^{j+1}(p)$ by dividing all intervals in half. Values, observations, and deviations are transferred to the new entries and the former entry is replaced. Because of the lack of further knowledge about the distribution of the values and observations, the number of observations and the deviation are equally divided to the new entries and the values remain.

Approximate value iteration allows comprehensive simulations of problem realizations and therefore a strong consideration of decision space, and transition space. Since the complexity of SDVRPs can often only be captured by extensive simulation, AVI is especially suitable for SDVPRs. Still, AVI requires the storage of values. Hence, an aggregation and a partitioning of the state space are generally necessary to apply AVI for SDVRPs. Here, the dynamic lookup table allows an subsequent adaption of the partitioning to the approximation process.

6.6 Hybrid Reactive Explicit Anticipation

In this section, we propose to combine online and offline reactive anticipatory approaches to exploit the advantages and to mitigate the shortcomings of the single ADP methods as described in Ulmer et al. (2017a). Powell (2011, p. 241ff) sketches the possibility of combining two different reactive anticipatory approaches to a *hybrid*

Table 6.4 Online versus offline

	Online	Offline
View	Local	Global
State	Detailed	Aggregated
Realizations	Few	Many
Decision points	Few	Many
Decision space	Limited	Comprehensive

reactive approach. First, we motivate the combination of online and offline anticipation in Sect. 6.6.1 by analyzing advantages and disadvantages of offline and online anticipation. We define a hybrid RA in Sect. 6.6.2 and show the benefits of the hybrid RA in a computational example in Sect. 6.6.3.

6.6.1 Motivation

As shown in Sects. 6.3 and 6.5, online and offline approaches have advantages and disadvantages depicted in Table 6.4. Online approaches are able to consider all features of a specific local state realization but not the global SDP-behavior. Offline approaches anticipate the global SDP but are usually not able to consider specific local state realizations in every feature. Online approaches allow a detailed state consideration. Nevertheless, online calculation time is limited. Hence, only a small number of realizations and decision points is considered. Further, the base policy limits the consideration of the decision space in the according SDP significantly and strongly impacts the overall reward in each realization. For a post-decision RA, given state S_k, m decisions, and n samples, the base policy is applied $n \times m \times (K - 1) - k$ times within the resulting SDP. Offline approaches allow considering many realizations, decision points, and decisions in the SDP because calculation time is unlimited. Still, the values have to be stored. As a result, an aggregation and a partitioning of the state space are generally necessary. This may result in a loss of solution quality because heterogeneous states are evaluated equally.

6.6.2 Hybrid Rollout Algorithm

In the following, we recall the proposal of Ulmer et al. (2017a) to combine online and offline algorithms by using an offline anticipatory base policy. This allows combining the advantages, alleviating the disadvantages of offline and online reactive anticipation and improving the RA's decision making. The *hybrid RA* combines two reactive anticipatory approaches and divides the computational challenge of solving the reduced SDP into learning and execution phase by using values derived by

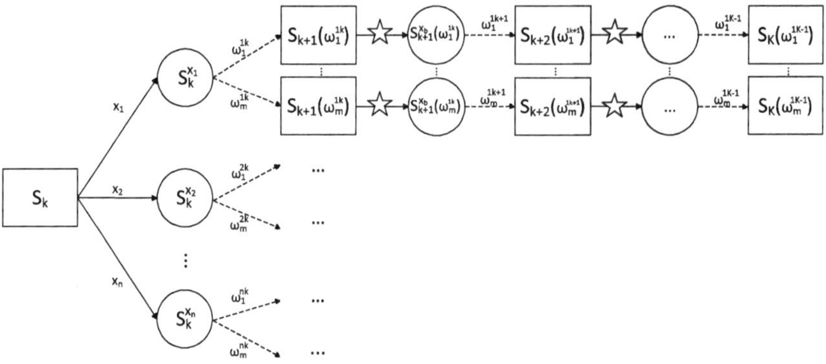

Fig. 6.5 A hybrid post-decision rollout

VFA (e.g., AVI) in an online RA. More specific, a hybrid RA uses the offline reactive approach as base policy. An exemplary hybrid post-decision RA is depicted in Fig. 6.5. At decision points, the hybrid RA draws on the solution of the offline SDP indicated by the stars.

In every decision point within the hybrid RA, the solution of the (aggregated) SDP provided by the offline approach is applied. The states' high level of detail in every decision point is maintained and the global SDP is considered in more detail regarding decisions and realizations. Following the analogy of model and pattern in Hand et al. (2001, pp. 165), the hybrid RA is able to consider a specific pattern (state) in every feature while keeping in mind the overall global model (SDP).

Compared to an online RA, hybrid anticipation can be achieved without the requirement of additional calculation effort in the execution phase. Compared to a plain online RA, a hybrid RA may be able to reduce the number of sample runs significantly or increase solution quality for the same number of sample runs respectively. A hybrid RA can also draw on IZS to further reduce calculation time.

6.6.3 *Example: Comparison of Online and Hybrid RAs*

In the following, we show the benefits of a hybrid RA. For this purpose, we present a small dynamic decision problem and compare a hybrid RA with a plain online RA. The hybrid RA draws on an offline reactive anticipatory base policy. The online RA draws on a myopic base policy.

Markov Decision Process

We consider a binary decision problem with only two different states per point of time (e.g., "on" and "off"). The according decision tree is depicted in Fig. 6.6.

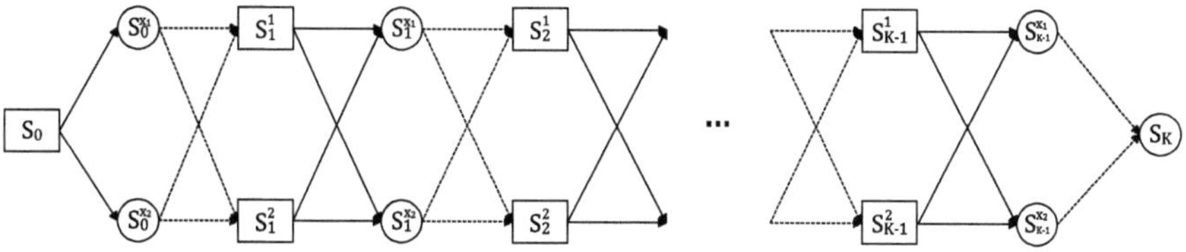

Fig. 6.6 Example for a stochastic dynamic program

The initial state is S_0 and can be associated with S_0^2. We set $K = 10$. In each decision point k, two possible decisions x_1, x_2 are given. x_1 leads to PDS $S_k^{x_1}$, x_2 leads to PDS $S_k^{x_2}$. The according rewards and transition probabilities are shown in Table 6.5. The rewards are $R(S_k^1, x_1) = 7.0$, $R(S_k^1, x_2) = 8.0$, $R(S_k^2, x_1) = 0.0$, $R(S_k^2, x_2) = 5.0$. The transition probabilities are $\mathbb{P}(S_k^{x_2}, S_{k+1}^2) = \mathbb{P}(S_k^{x_1}, S_{k+1}^1) = 0.9$ and $\mathbb{P}(S_k^{x_1}, S_{k+1}^2) = \mathbb{P}(S_k^{x_2}, S_{k+1}^1) = 0.1$ for $k \leq K - 2$. The transition probabilities reflect a relatively stable system. In 90% of the cases, PDS $S_k^{x_i}$ results in state S_{k+1}^i for $i = 1, 2$.

In this example, we only examine the first decision in decision point $k = 0$ and state S_0^2. In this decision point, the decision is about turning the system on (x_1) or leaving it switched off (x_2). Turning the system on results in no immediate reward $R(S_0^2, x_1) = 0.0$. Leaving it switched off results in a reward of $R(S_0^2, x_2) = 5.0$.

Optimal Decision Policy

The optimal decision policy π^* can be calculated by solving the SDP. The according decision rules $X_k^{\pi^*}$ for state S_k^j are depicted in Eq. 6.18. The decision in the first six decision points is x_1. From decision point $k = 6$ to $k = 9$, the decision depends on the current state. The decision in the last decision point is x_2. The expected value of S_0 is $V(S_0) = 58.6$.

$$X_k^{\pi^*}(S_k^j) = \begin{cases} x_1 : 0 \leq k < 6 \\ x_j : 6 \leq k < 9 \\ x_2 : \quad\quad k = 9 \end{cases} \tag{6.18}$$

As a result, the decision x_1 in S_0^2 to turn the system on is optimal. Compared to the optimal decision policy π^*, we analyze the decision making of a κ-step RA by varying the parameter κ. Since the SDP for this problem is efficiently solvable, we do not sample within the κ-step RA, but consider the transition probabilities $\mathbb{P}$. We compare two base policies. The first base policy π_m is myopic resulting in an online RA. The second base policy π_a is reactive anticipatory resulting in a hybrid RA.

Table 6.5 Transition probabilities and rewards

(PD) State	S_k^1	S_k^2	S_k^1	S_k^2
$S_k^{x_1}$	0.9	0.1	7.0	0.0
$S_k^{x_2}$	0.1	0.9	8.0	5.0

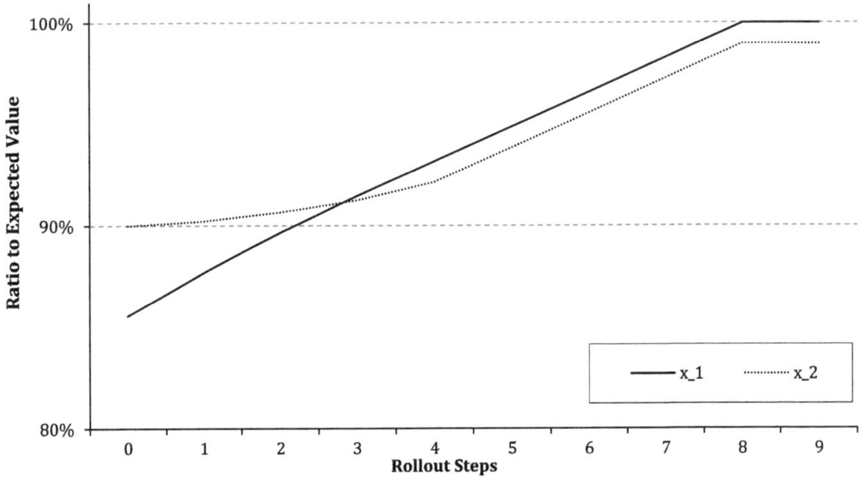

Fig. 6.7 Development of expected values for *myopic*

Online Rollout

The first base policy π_m we apply selects always the decision x_2 leading to the highest immediate reward as depicted in Eq. 6.19. The application of this base policy results in an online κ-step RA.

$$X_k^{\pi_m}(S_k^j) = x_2 : 0 \leq k \leq 9 \tag{6.19}$$

We consider two special cases: $\kappa = 0$ and $\kappa = K - 1$. $\kappa = K - 1$ solves the SDP perfectly resulting in policy π^* and the selection of the optimal decision x_1 and an expected value $V(S_0^{x_1}) = 58.6$. $\kappa = 0$ results in a post-decision RA. Combined with π_m, this leads to expected PDS values of $\hat{V}(S_0^{x_1}) = 50.1$ and $\hat{V}(S_0^{x_2}) = 47.7$. Because x_2 results in an immediate reward of $R(S_0, x_2) = 5$ while $R(S_0, x_1) = 0$, the post-decision RA with base policy π_m selects decision x_2.

Starting with $\kappa = 0$, we now increase κ to analyze when the online RA is able to identify the same decision as the optimal policy π^*. For every κ, we calculate the immediate reward $R(S_0, x_j)$ plus the RA's outcome $\hat{V}(S_0^{x_j})$ for the two decisions x_j, $j = 1, 2$. Figure 6.7 shows the ratio of $R(x_j) + \hat{V}(S_0^{x_j})$ to the optimal value $V(S_0)$. On the x-axis, the number of RA steps κ is depicted. We already described

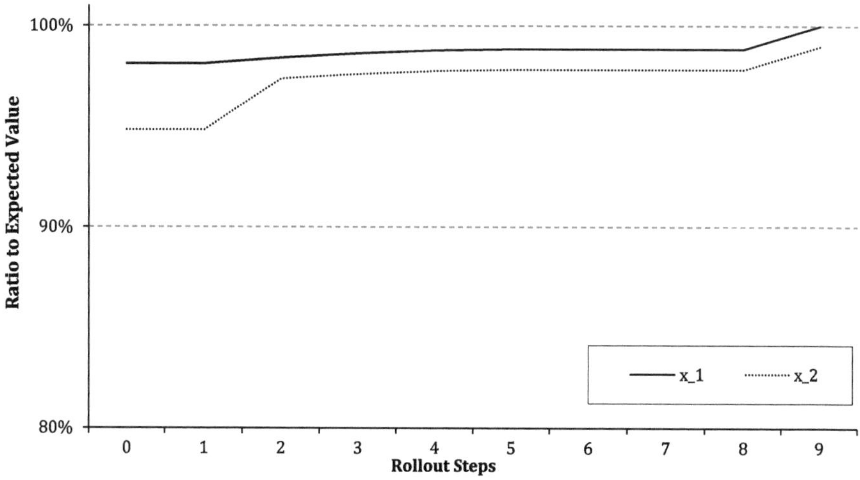

Fig. 6.8 Development of expected values for *ahead*

$\kappa = 0$. For $\kappa = 9$, the exact value is reached. As we can see, for $\kappa \geq 3$, decision x_1 is selected. Even though the RA's values are still different to the exact values, the PDSs' values are already in correct order allowing the selection of an optimal decision. For this example with base policy π_m, a 3-step RA would provide the optimal decision in S_0.

Hybrid Rollout

The second base policy π_a we apply is offline reactive anticipatory and is called *ahead*. *ahead* can be seen as a lookahead for one following decision point. The decision rules for π_a consider the immediate reward and the expected maximal reward in the next decision point. Let $p = 3 - i$ be the indicator for the alternative to i, i.e., $i = 1 \Rightarrow p = 2, i = 2 \Rightarrow p = 1$. Given state S_k^j with $k < 9$, this results in the selection of decision $X_k^{\pi_a}(S_k^j)$ as depicted in Eq. 6.20.

$$X_k^{\pi_a}(S_k^j) = \underset{x \in \{x_1, x_2\}}{\arg\max} \left\{ R(S_k^j, x) + 0.9 \max_{x \in \{x_1, x_2\}} \left\{ R(S_{k+1}^i, x) \right\} + 0.1 \max_{x_i \in \{x_1, x_2\}} \left\{ R(S_{k+1}^p, x) \right\} \right\} \tag{6.20}$$

The according decision policy is depicted in Eq. 6.21. Notably, the plain application of the offline policy π_a in S_0 leads to the suboptimal decision x_2. The RA drawing on *ahead* is offline and online anticipatory and therefore hybrid.

$$X_k^{\pi_a}(S_k^j) = \begin{cases} x_k^j : 0 \leq k < 9 \\ x_k^2 : \quad\quad k = 9 \end{cases} \tag{6.21}$$

Again, we vary κ between $\kappa = 0$ and $\kappa = 9$. The according development of the expected values for x_1 and x_2 is depicted in Fig. 6.8. Even though the values $\hat{V}$ are still smaller than the exact values, the tendency is correct for all κ. The post-decision RA with $\kappa = 0$ already provides the optimal decision in S_0.

Interpretation

This example emphasizes the importance of a suitable base policy selection, especially for post-decision RAs. The use of the hybrid RA with anticipatory base policy π_a provides the optimal decision already in a post-decision RA, i.e., significantly earlier than the online RA drawing on π_m. For SDVRPs, the number of possible realizations and transitions is usually vast. As a result, only post-decision RAs can be applied in many cases. The combination of online and offline reactive anticipation strengthens the solved SDP and the according approximation without additional online calculation time and results in an improvement of the decision policy achieved by a post-decision RA.

A hybrid RA allows the consideration of both a particular state in detail and the overall object system in general. While the close environment of a state in the SDP is captured by the online simulation of the RA in high detail, the overall system behavior is integrated via the anticipatory base policy derived by AVI. This base policy significantly alleviates the RA's shortcoming of a limited decision and transition space consideration. Notably, the offline AVI-values may also be used to limit the online simulation horizon (Ulmer and Hennig 2016).

Chapter 7
Literature Classification

In Chap. 3, we have derived the four main drivers of uncertainty in practical routing applications. In Chap. 4, we have defined and modeled stochastic dynamic vehicle routing problems. In Chaps. 5 and 6, we have presented a classification of anticipation and an overview of anticipatory solution approaches. We now use the findings of these previous chapters to classify work on SDVRPs regarding uncertainty, modeling, and the applied (anticipatory) solution approaches. There is a number of literature reviews classifying SDVRPs, e.g., by Gendreau and Potvin (1998); Bianchi (2000); Ghiani et al. (2003); Larsen et al. (2008); Pillac et al. (2013), or Ritzinger et al. (2015). Most recently, Psaraftis et al. (2015) have given an extensive overview. The reviews mainly list works chronologically. A literature review with focus on modeling and anticipation is not provided yet. In the following, we present an efficient classification of the most prominent work regarding SDVRPs. We differentiate problems regarding the four drivers of uncertainty. For each driver, we focus on the objective, the modeling of problem and uncertainty, and classify the solution approaches regarding the degree of anticipation. For the approaches providing reactive anticipation, we also analyze the point of time the value determination is conducted, i.e., online or offline. The main purpose of the classification is to analyze the chronological process for the different problem classes regarding modeling and anticipation as well as to highlight the most common reactive anticipatory approaches applied to SDVRPs. We mainly focus on papers listed in Psaraftis et al. (2015). We include additional papers providing reactive anticipation as well as papers with static models and recourse actions. The classification in detail is presented in Sect. 7.1. The classification of SDVRPs for stochastic travel time is given in Sect. 7.2, for service time in Sect. 7.3, for demands in Sect. 7.4, and for requests in Sect. 7.5. We conclude this chapter with an analysis of the developments of modeling and solution approaches over time in Sect. 7.6.

© Springer International Publishing AG 2017

M.W. Ulmer, *Approximate Dynamic Programming for Dynamic Vehicle Routing*,
Operations Research/Computer Science Interfaces Series 61,
DOI 10.1007/978-3-319-55511-9_7

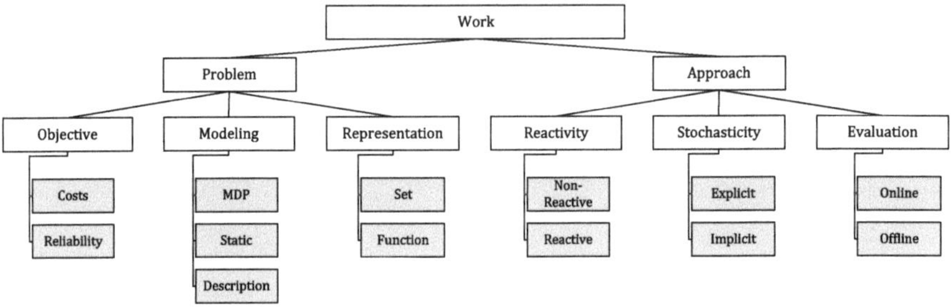

Fig. 7.1 Literature classification

7.1 Classification

For every driver of uncertainty, we classify the according problems by the classification depicted in Fig. 7.1. We classify the presented problems regarding objective, modeling, and the representation of uncertainty. We further depict the applied approaches and the degree of anticipation as defined in Chap. 5.

For the problem, we differentiate objectives as maximizing reliability (r) and minimizing required costs (c). In this context, costs are mainly represented by the travel time or distance and the number of vehicles, while reliability is represented by the number of demands and customers served and the compliance of time limit and deadlines. For the modeling, we differentiate between modeling as a MDP (m), modeling as static by a mixed integer program with dynamic adaptions (or recourse actions) (s), and by a (textual) description (d). For modeling as an MDP, we distinguish the sole description as MDP (m(d)) and the explicit mentioning of an MDP (m(e)). We differentiate the representation ("Rep.") of uncertainty in the model or if not mentioned explicitly, in the definition of the instances. As representation, we distinguish stochastic functions (f) and sets of possible realizations (s). We classify approaches as reactive (r) or non-reactive (-) anticipatory. Further, we differentiate between implicit (i) and explicit (e) use of stochastic information. If the approach is reactive, we depict when the (approximate) evaluation of states is conducted (Eval.), online (n) or offline (f). Notably, also non-reactive approaches may require extensive calculation time.

7.2 Travel Times

In many dynamic vehicle routing problems, travel times are modeled as deterministic, but time-dependent (Malandraki and Daskin 1992; Barceló et al. 2007; Van Woensel et al. 2008; Respen et al. 2014). In other cases, travel times are stochastic, but planning is conducted statically a priori (Laporte et al. 1992; Zielinski 2004; Montemanni and Gambardella 2004; Montemanni et al. 2007; Lecluyse et al. 2009; Campbell et al.

Table 7.1 Stochastic travel times

	Problem setting			Anticipation		
	Objective	Modeling	Rep.	Reactive	Stochastic	Eval.
Xiang et al. (2008)	r,c	d	f	-	i	-
Lorini et al. (2011)	r,c	d	f	-	i	-
Ghannadpour et al. (2013)	r,c	d,s	f	-	i	-
Yan et al. (2013)	r,c	d,s	s	r	e	n
Schilde et al. (2014)	r,c	d	f	r	e	n
Ehmke and Campbell (2014)	r,c	d	f	r	e	n
Köster et al. (2015)	c	d	s,f	-	i	-

2011; Han et al. 2013; Adulyasak and Jaillet 2015; Ehmke et al. 2015; Groß et al. 2015; Ehmke et al. 2016). As a result, the number of SDVRPs with stochastic travel times is limited as shown in Table 7.1.

In most of the cases, the objectives of the presented problems consider both reliability and travel costs. The problems are modeled by descriptions in all cases. In some cases, a static mixed integer program is provided. Uncertain travel times are modeled as normally (Xiang et al. 2008; Lorini et al. 2011), triangularly (Ghannadpour et al. 2013), or Burr-distributed (Ehmke and Campbell 2014). Yan et al. (2013) model uncertainty by a finite set of possible realizations. Schilde et al. (2014) model travel times as dependent travel speed. The travel speed changes due to stochastic occurrences of congestion in certain areas. Köster et al. (2015) model uncertainty in travel times by a set of possible travel time matrices with stochastic transitions between the matrices. They combine the resulting travel time with additional normally distributed noise.

Xiang et al. (2008); Ghannadpour et al. (2013); Lorini et al. (2011); Köster et al. (2015) re-optimize regarding the current information in every decision point. Reactive anticipatory approaches are applied by Yan et al. (2013); Schilde et al. (2014); Ehmke and Campbell (2014). Yan et al. (2013) calculate the expected penalties for violating time windows online and adapt their planning accordingly. The calculation is possible due to a relatively small number of possible realizations. Schilde et al. (2014) sample future occurrences of congestion and evaluate plans in a dynamic stochastic variable neighborhood search (Schilde et al. 2011), a procedure related to RAs. Ehmke and Campbell (2014) dynamically decide about the acceptance of dynamic requests before the actual routing is conducted. They approximate the standard deviation of the arrival time for every customer by assuming normally distributed travel times. They then determine a latest arrival time for every customer. If the planned arrival time exceeds the latest arrival time, the customer is not accepted.

Table 7.2 Stochastic service times

	Problem setting			Anticipation		
	Objective	Modeling	Rep.	Reactive	Stochastic	Eval.
Delage (2010)	r	d	f	-	i	-
Maxwell et al. (2010)	r	m(e)	f	r	e	f
Schmid (2012)	r	m(d)	f	r	e	f
Zhang et al. (2015a)	r	m(e)	f	r	e	f

7.3 Service Times

Stochastic service times are often considered given static problems (Laporte et al. 1992; Hadjiconstantinou and Roberts 2001; Kenyon and Morton 2003; Tang and Miller-Hooks 2005; Jula et al. 2006; Li et al. 2010; Sungur et al. 2010; Campbell et al. 2011; Souyris et al. 2013; Zhang et al. 2014). The work on dynamic vehicle routing problems with stochastic service times is limited. In their review, Psaraftis et al. (2015) only mention a single dynamic vehicle routing problem with stochastic service times presented by Errico et al. (2013). The considered problem is defined as static. Subsequent work of Errico et al. (2016) as well as work of Yuan et al. (2015) still model the problem as static, but allow recourse actions if certain stochastic events occur. As mentioned earlier, a static problem with recourse actions can be associated with an a priori definition of a policy for a dynamic problem. Work specifically considering a dynamic vehicle routing problem with stochastic service times is conducted by Delage (2010); Maxwell et al. (2010); Schmid (2012), and Zhang et al. (2015a) depicted in Table 7.2. Delage uses a re-optimizing algorithm to route a fleet of technicians given uncertain service times. Maxwell et al. (2010) and Schmid (2012) consider ambulance dispatching problems, where requests and service times are uncertain. Maxwell et al. (2010) uses MVF to approximate expected future penalties. Service times are exponentially distributed. Schmid (2012) uses approximate value iteration to derive a reactive decision policy explicitly anticipating future requests and service times. Service times are represented by a Gamma distribution derived from historical observations. Zhang et al. (2015a) consider an orienteering problem (compare Golden et al. 1987) with uncertain waiting and service times respectively. They model service (waiting) times via queueing and use RAs for reactive anticipation.

7.4 Demands

In the following, we analyze SDVRPs with stochastic demands depicted in Table 7.3. The objective for SDVRPs with stochastic demands is generally to fulfill customer demands at minimal costs. Most of the work considers the distribution of goods to

Table 7.3 Stochastic demands

	Problem setting			Anticipation		
	Objective	Modeling	Rep.	Reactive	Stochastic	Eval.
Secomandi (2000)	c	m(e)	s	r	e	f
Secomandi (2001)	c	m(e)	s	r	e	n
Christiansen and Lysgaard (2007)	c	d,s	f	r	e	f
Novoa and Storer (2009)	c	m(e)	s	r	e	n
Secomandi and Margot (2009)	c	m(e)	s	r	e	f
Pandelis et al. (2012)	c	d	f	r	e	f
Côté et al. (2013)	c	d,s	s	r	e	f
Goodson et al. (2013a)	c	m(e)	f	r	e	n
Goodson et al. (2013b)	c	m(e)	f	r	e	n
Coelho et al. (2014)	c	m(d)	s	r	e	n
Brinkmann et al. (2015)	r	m(e)	f	r	e	f
Sarasola et al. (2015)	c	d,s	f	r	e	n

a set of customers given vehicle capacities. Coelho et al. (2014); Brinkmann et al. (2015) consider inventory routing problems. Coelho et al. (2014) use sampling of customer demands within a lookahead to achieve reactive anticipation. Brinkmann et al. (2015) determine suitable time-dependent fill levels regarding average future demands. Brinkmann et al. (2015) maximize the fulfilled customer demands regarding a limited fleet.

Notably, all approaches allow reactive anticipation and many problems are modeled as MDPs. Christiansen and Lysgaard (2007); Côté et al. (2013) model the problem as static with recourse actions resulting in dynamic adaptions of the plan. As anticipatory approaches, RAs (Secomandi 2001; Novoa and Storer 2009; Goodson et al. 2013a, b) and VFAs (Secomandi 2000; Secomandi and Margot 2009) are applied. Sarasola et al. (2015) apply a stochastic and dynamic variable neighborhood search sampling future demands. We assume that the high percentage of reactive anticipation is enabled by the generally reasonably sized state space. For many problems, the number of customers is small (Pandelis et al. 2012). In some cases, the sequence or some sub-sequences of customers are predefined (Secomandi and Margot 2009).

7.5 Requests

Compared to the other drivers of uncertainty, the work on SDVRPs with stochastic requests is relatively comprehensive. The work mainly considers service- or pickup-requests (e.g., Bent and Van Hentenryck 2004; Thomas 2007). In some cases, also pickup and delivery is considered (e.g., Psaraftis 1980; Savelsbergh and Sol 1998; Sáez et al. 2008). The classification is depicted in Table 7.4.

Table 7.4 Stochastic requests

	Problem setting			Anticipation		
	Objective	Modeling	Rep.	Reactive	Stochastic	Eval.
Psaraftis (1980)	c	m(d)	f	-	i	-
Powell et al. (1988)	r	m(d)	s	r	e	f
Bertsimas and Van Ryzin (1991)	c	d	f	-	i	-
Papastavrou (1996)	c	d	f	-	i	-
Tassiulas (1996)	c	d	f	-	i	-
Savelsbergh and Sol (1998)	c	d,s	f	-	i	-
Gendreau et al. (1999)	r,c	d,s	f	-	i	-
Swihart and Papastavrou (1999)	r	d	f	-	i	-
Ichoua et al. (2000)	r, c	d,s	f	-	i	-
Larsen et al. (2002)	c	d	f	-	i	-
Mitrović-Minić and Laporte (2004)	c	d,s	f	-	i	-
Thomas and White III (2004)	r,c	m(e)	s	r	e	f
Bent and Van Hentenryck (2004)	r	d	f	-	e	-
Van Hemert and La Poutré (2004)	r	d	s	r	e	f
Branke et al. (2005)	r	d	f	r	e	f
Ichoua et al. (2006)	r,c	m(e),s	f	r	e	n
Gendreau et al. (2006)	r,c	d,s	f	-	i	-
Hvattum et al. (2006)	r,c	d,s	f	-	e	-
Chen and Xu (2006)	c	d,s	s	-	i	-
Thomas (2007)	r	m(e)	s	-	e	-
Bent and Van Hentenryck (2007)	r	d,s	f	-	e	-
Hvattum et al. (2007)	r	d,s	f	-	e	-
Sáez et al. (2008)	c	m(d)	f	r	e	f
Pureza and Laporte (2008)	r,c	d	f	-	i	-
Ghiani et al. (2009)	r	d,s	f	r	e	n
Angelelli et al. (2009)	r	d	f	-	i	-
Maxwell et al. (2010)	r	m(e)	f	r	e	f
Mes et al. (2010)	r	d	f	r	e	n
Meisel et al. (2011)	r	m(d)	s	r	e	f
Schmid (2012)	r	m(d)	f	r	e	f
Ghiani et al. (2012)	r	m(e)	s	-	e	-
Sheridan et al. (2013)	r	d	f	-	i	-
Ulmer et al. (2015a)	r	d,s	f	r	e	f
Ulmer et al. (2017b)	r	m(e)	f	r	e	f
Ulmer et al. (2015b)	r	m(e)	f	r	e	n

The objective in most of the considered problems is to provide reliable service. Mainly, objective functions aiming on reliability consider the violations of time windows (Bent and Van Hentenryck 2004), the inconvenience (or system time) of a customer (Ghiani et al. 2009; Schmid 2012; Sheridan et al. 2013), or the overall number of services (Branke et al. 2005; Thomas 2007). Objective functions aiming on costs minimize travel distance (Psaraftis 1980) or the number of vehicles and travel time (Hvattum et al. 2006). Some objective functions combine reliability and costs. As an example, Gendreau et al. (2006) minimize travel time and time window violations. In contrast to the other drivers of uncertainty, non-reactive explicit anticipatory approaches are applied. Thomas (2007) uses request probabilities of potential customers to calculate effective waiting points. Bent and Van Hentenryck (2004); Hvattum et al. (2006); Bent and Van Hentenryck (2007), and Hvattum et al. (2007) sample requests based on the spatial request distribution to compare plans using a consensus function. Reactive anticipatory approaches are manifold. Van Hemert and La Poutré (2004) evaluate post-decision states by the amount of feasible visits of *fruitful regions* with high request probabilities. Branke et al. (2005) evaluate plans regarding the coverage of the service area and the resulting probability to serve a future request. Ghiani et al. (2009) sample customers in the short term future to estimate future time window violations. Meisel et al. (2011) apply AVI without aggregation or partitioning for a problem with only a small number of potential customers. Ulmer et al. (2017b) apply AVI aggregating PDSs to vectors of time and free time budget. Ulmer et al. (2015b) apply an RA by sampling future requests regarding the spatial-temporal customer distribution.

7.6 Analysis

In this section, we analyze the developments in modeling and anticipation over the years. Therefore, we draw on the work classified earlier. Since the classification focuses on (reactive) anticipation, the selection of the work may be biased. Nevertheless, the presented tendencies can be generalized. First, we analyze the papers regarding the time of publication. We then show the development of the problems' objective function and modeling and analyze the approaches regarding reactivity and the inclusion of stochasticity.

7.6.1 Time Distribution

Figure 7.2 shows the time distribution of the papers considered in Tables 7.1, 7.2, 7.3 and 7.4. On the x-axis, the years are shown. On the y-axis, the number of publications within the time span is depicted. The general trend follows the development shown in FIG. 1. of Psaraftis et al. (2015). Besides some theoretical pioneer work in the 1980 s by Psaraftis (1980) and Powell et al. (1988), the work on SDVRPs for practical applications started in the 1990 s and is increasing since then. In the following, we focus on the 53 referenced papers published since 1991.

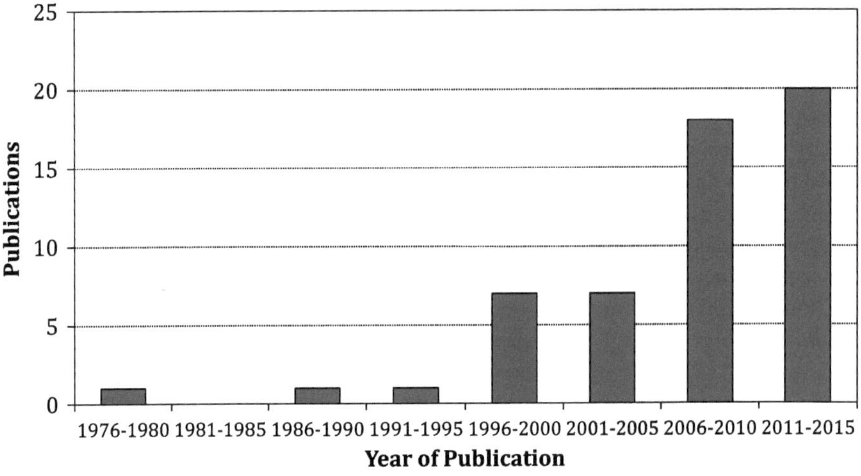

Fig. 7.2 Time distribution of the referenced publications

7.6.2 *Problem*

In this section, we analyze the problems' objectives and modeling. First, we show how the objective function has developed over the years. Then, we differentiate how the problem is modeled.

Objective Function

Figure 7.3 shows the development of the objectives *costs* and *reliability* over time. On the x-axis, the years are depicted. On the y-axis, the percentage of work on problems with the according objective is shown. Notably, some objectives contain both costs and reliability. A constant increase of work considering reliability can be seen. The amount of work on problems with costs as an objective is decreasing over the years. Practical applications inducing SDVRPs are usually located on the operational planning level. On the operational level, resources like working hours or the fleet of vehicles are generally given and the objective is to use these resources to provide effective customer services and to avoid customer inconveniences.

Modeling

We now analyze how the problems are modeled. We differentiate between modeling by description, modeling as static (by a mixed integer program), or modeling as MDP. Some problems are modeled in several ways. Figure 7.4 shows the percentage of modeling techniques over time. Initially, the problems were described and not

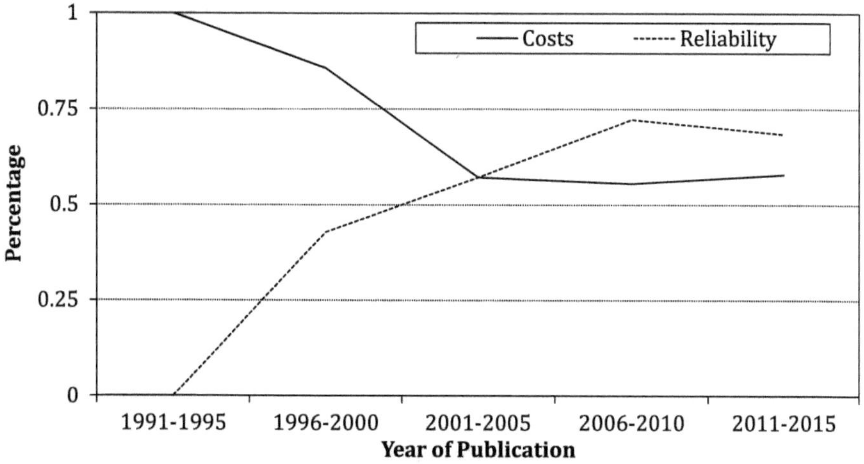

Fig. 7.3 Objectives of the referenced problems

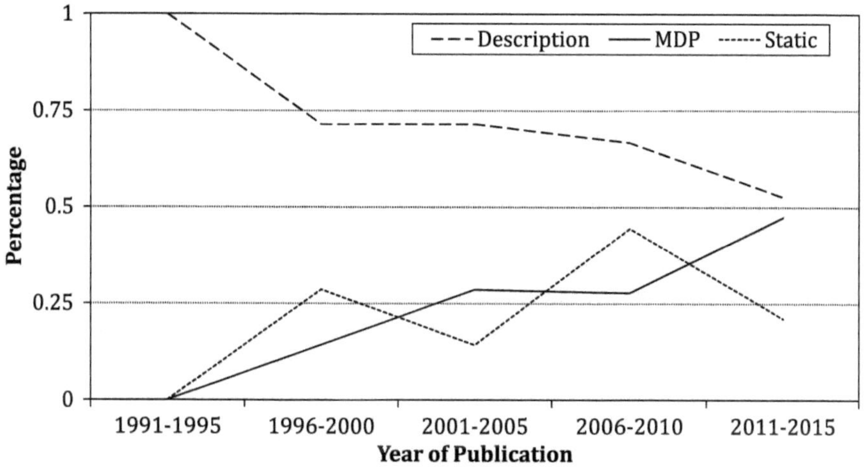

Fig. 7.4 Modeling of the referenced problems

scientifically modeled. Since then, we experience a constant increase in work modeling problems as MDPs while the percentage of modeling by description decreases. The use of static mixed integer programs varies over time. The number of work modeling SDVRPs as static problems seems to decrease. Nevertheless, for SDVRPs represented by static problems with recourse actions, it is the most common way of modeling.

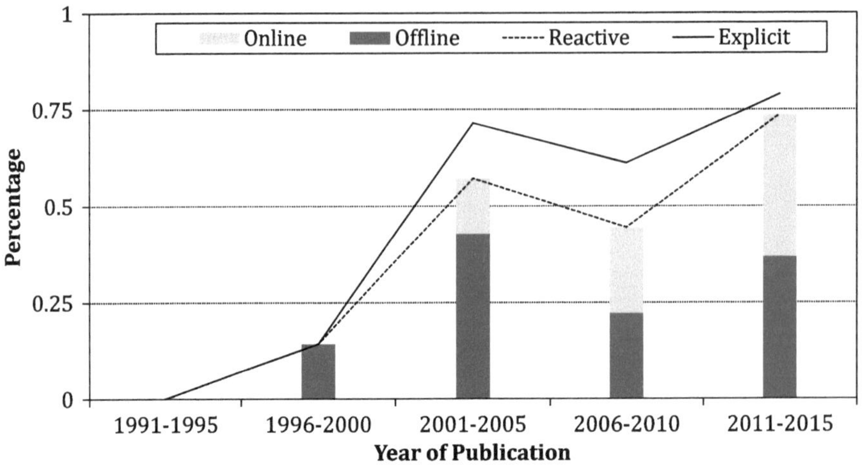

Fig. 7.5 Anticipation of the referenced approaches

7.6.3 Approaches

In the following, we analyze the applied approaches. We differentiate approaches regarding reactivity and explicitness in anticipation as shown in Fig. 7.5. We further show the portions of online and offline reactive anticipation. At first, we look at the development of explicit approaches including stochastic information. We experience a constant and significant increase over the years. Between 2011 and 2015, 79.0% of the approaches allowed the integration of stochastic information. The same behavior can be seen for the development of reactive anticipatory approaches. The gap between the two curves is caused by the non-reactive explicit anticipatory approaches, e.g., by Bent and Van Hentenryck (2004) or Thomas (2007). We are not aware of any reactive implicit anticipatory approaches for SDVRPs. The first reactive anticipatory approaches were offline. With an increase of computational resources, the amount of online approaches increased. Since 2006, the amount of offline and online reactive anticipatory approaches is balanced.

7.7 Implications

This literature review closes the first part of this book. As a conclusion of the first part and the literature review, we can state that the importance of practical routing applications with uncertainty and the requirement for frequent replanning as well as the work on SDVRPs is constantly growing. Data analysis and computational resources allow the integration of stochastic information and future events into decision making, i.e., reactive explicit anticipation.

The work on SDVRPs and the applied approaches vary for the different drivers of uncertainty. While for stochastic demands and requests, reactive explicit anticipation and the according terminology are established and applied in a majority of the publications, work on SDVRPs and reactive anticipation is rare for stochastic service and travel times. The reasons for this development may be twofold. On the one hand, compared to requests or demands, the impact of service times and travel times may be less significant on planning and the according SDVRPs do not require replanning. Hence, dynamic vehicle routing and anticipation may not be necessary for some applications. On the other hand, the derivation of (reactive) anticipatory approaches for stochastic service and travel times may be challenging, e.g., compared to SDVRPs with stochastic demands. In the future, due to increasing urbanization, research on SDVRPs with stochastic travel times may be worthwhile. Due to the demographic developments, the increasing requirement for patient transport and physician routing may demand a more comprehensive consideration of stochastic service times in SDVRPs. Uncertainty in demands may decrease because of information technologies. Eventually, the first part of this book revealed stochastic requests as the major challenge for dispatchers. Therefore, we will analyze an SDVRP with stochastic requests in the second part of this book.

Part II
Stochastic Customer Requests

Chapter 8
Motivation

In the second part of the book, we describe the process of modeling and the derivation of anticipatory solution approaches for a specific SDVRP. As we have seen in Chap. 3, the main reason for replanning is given by uncertain customer requests. Therefore, we consider a *dynamic vehicle routing problem with stochastic customer requests.* Uncertain requests occur in many different routing applications. In Chap. 3, we have identified uncertain customer requests occurring for service vehicle, (e.g., emergency vehicles and technicians), for passenger transport (e.g., dial a ride, transport of patients and elderly or shared mobility), and for transportation of goods (e.g., parcel pickup, courier services, and same-day delivery). All these applications are only able to conduct short-term or maximally medium-term planning of the vehicles' routes. Therefore, the requirement for replanning for these types of problems is high. We expect that the impact of anticipation on decision making and solution quality is significant. In the following, we motivate and present the particular practical routing application. An outline of the second part of this book is given in Sect. 8.3.

8.1 Application

We focus on parcel pickups. As described in Sect. 2.5, the parcel market is highly competitive and the requirement for short delivery times increases. Due to a significant increase in e-commerce sales, the number of shipped customer to customer (C2C) and business to customer (B2C) parcels has grown significantly. In Germany, the B2C and C2C market has increased about 50% in the last five years (Esser and Kurte 2015). Many small vendors use online market places to sell their products directly to the purchaser. For successful e-commerce, delivery times and delivery costs are two of the main influence factors (Lowe et al. 2014). Customers expect reasonably priced, fast, and reliable service (Ehmke 2012). About 20% of all parcels are shipped via courier or express delivery. Many of the parcels are picked up directly at

© Springer International Publishing AG 2017

M.W. Ulmer, *Approximate Dynamic Programming for Dynamic Vehicle Routing*,
Operations Research/Computer Science Interfaces Series 61,
DOI 10.1007/978-3-319-55511-9_8

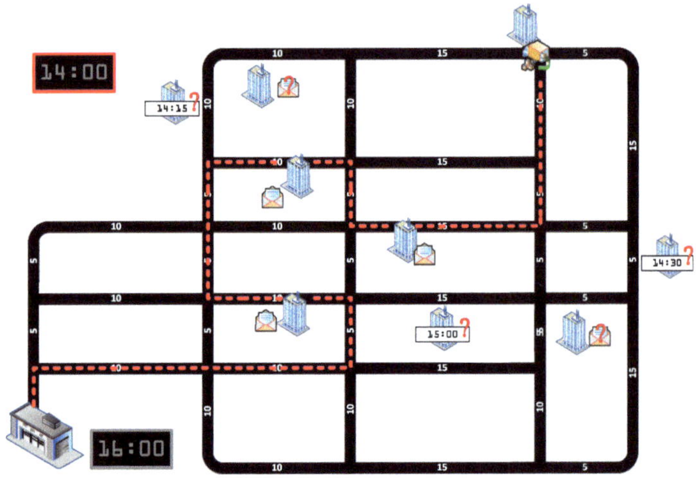

Fig. 8.1 Motivational example: planning situation

the seller (in the following called *customer*) and are processed the same day (Hvattum et al. 2006). Some requests are known in the beginning but most of these pickups are requested in the course of the day (Lin et al. 2010). To serve these requests, courier express and parcel services schedule vehicles dynamically. The collected parcels are then transported to the depot to be shipped long haul (Pillac et al. 2013). In urban transportation, drivers' wages are the main cost factor. Since the drivers are already contracted for a daily working time, driving costs can be viewed as fixed (Thomas 2007). The beginning and the end of the working hours define the *shift* and result in a *time budget* for serving requests. Usually, the early request customers have to be served. Since the time budget is limited, not all dynamic requests can be confirmed. For every request, the dispatcher has to decide about a confirmation or rejection (Gendreau et al. 1999). Rejected requests may cause penalties or future handling costs, since they may be served by a third party or postponed to following days (Angelelli et al. 2009). To include new requests, the dispatcher has to adapt the planned tour (Gendreau et al. 2006). The duration of the planned tour has to be feasible regarding the time limit. Service providers aim on a large number of confirmations to maximize revenue and purchasers' satisfaction and to reduce handling costs for the rejected requests.

Current decision making impacts the *expected number of future confirmations* (Bellman 1956). To allow maximizing the sum of immediate and future confirmations, future requests and future decisions have to be considered in current decisions. For anticipation of future requests, the service provider can derive stochastic probabilities of customer requests for certain regions of the service area by making prognoses about customer behavior (Dennis 2011). The requesting customers are arbitrarily distributed in the whole service area. Request times and locations are not known beforehand. As a result, anticipation is challenging.

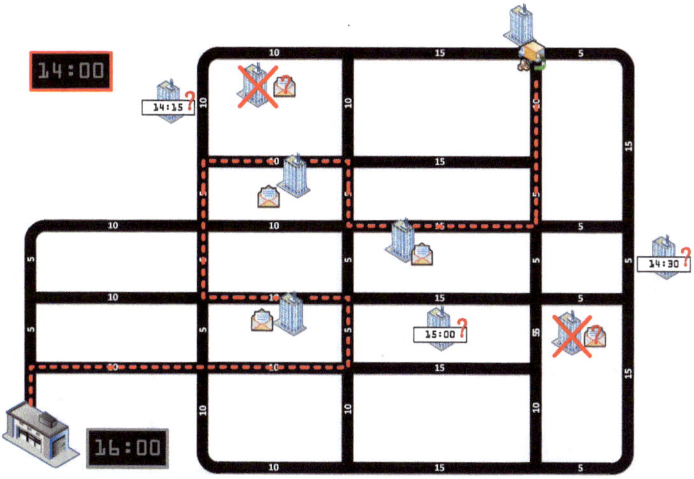

Fig. 8.2 Motivational example: plan 1

We focus on the uncertainty of customers' requests. Therefore, we consider a vanilla problem setting. We assume travel times to be known and certain and neglect service times and demands. For courier services, service times are usually not given and capacities are neglectable. We further focus on a single vehicle and assume requesting customers to be of equal priority. Still, priorities can be easily integrated in the approaches by adapting the rewards of confirmations.

8.2 Replanning and Anticipation

In the following, we present an exemplary planning situation for a dispatcher and describe both the requirement for replanning and the impact of anticipation. We highlight the requirement for spatial and temporal anticipation of requests. A planning situation occurs when exogenous request process and planning process are synchronized. Following the literature, requests are accumulates until the vehicle reaches a customer's location (Thomas 2007). Hence, a planning situation occurs when the vehicle has served a customer. In Fig. 8.1, the service area with an exemplary street network is depicted. The numbers on the segments indicate the travel time in minutes. We assume that turning within the segments is not possible. In the current planning situation, it is 2 p.m. in the afternoon. The vehicle just served the customer in the upper right part of the network. The depot is located in the lower left. The vehicle has to return to the depot at 4 p.m. to enable same-day long haul shipping. As a result, 2 h or 120 min of travel time are left. There are currently 3 confirmed customers the vehicle has to visit. A feasible tour is indicated by the dashed line. The tour allows

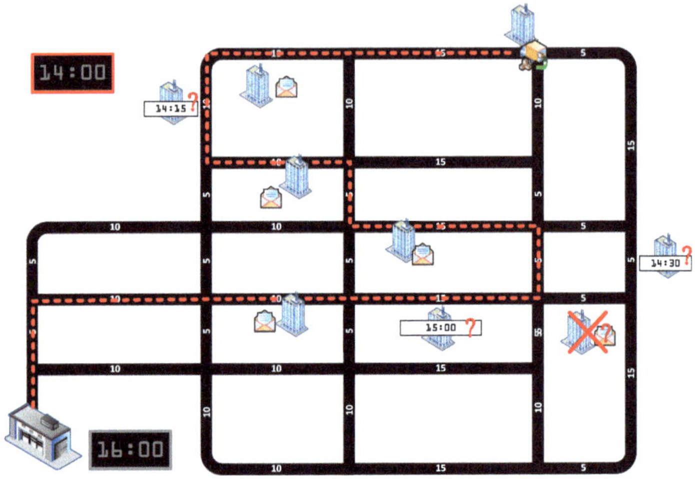

Fig. 8.3 Motivational example: plan 2

service of the 3 confirmed customers. The tour duration is 90 min. This tour would provide the dispatcher with 30 min of free time budget to include new requests. The dispatcher has to answer 2 new requests, indicated by the question marks on the envelopes. One request is located in the upper left corner, one is located in the lower right corner of the service area. Further, there are three future requests, which are unknown for the dispatcher, but are shown for the purpose of presentation. These future requests and the requests times are indicated by the time signs and question marks.

In this planning situation, the dispatcher has to decide about a confirmation or rejection for each of the new requests and about the next customer to visit. We assume, that no diversion is possible, i.e., the vehicle has to serve the next customer first regardless new requests. Because of the time limit, the confirmation of both requests is not possible. Three possible plans remain. In Figs. 8.2, 8.3 and 8.4, we show the three plans by confirmation of only the request located in the bottom right, only the request located in the top left, and none of the requests.

In the first plan, depicted in Fig. 8.2, none of the requests is confirmed and the planned tour is maintained. 30 min of the time budget are still free to confirm later requests. This flexibility allows the dispatcher to extend the tour at any point for the inclusion of future requests. As a result, the dispatcher will be able to confirm the upper left customer request in the future.

In Fig. 8.3, the second plan is shown. Only the request in the bottom right is confirmed. The dashed line indicates the planned tour with a duration of 110 min, only 10 min remain. As we can see, the planned tour is an adaption of the previous feasible tour. The next customer to be visited is the confirmed request. We now examine the impact of the plan on the three future requests. Because of the time limit, the vehicle is not able to include the upper left request and the request in the

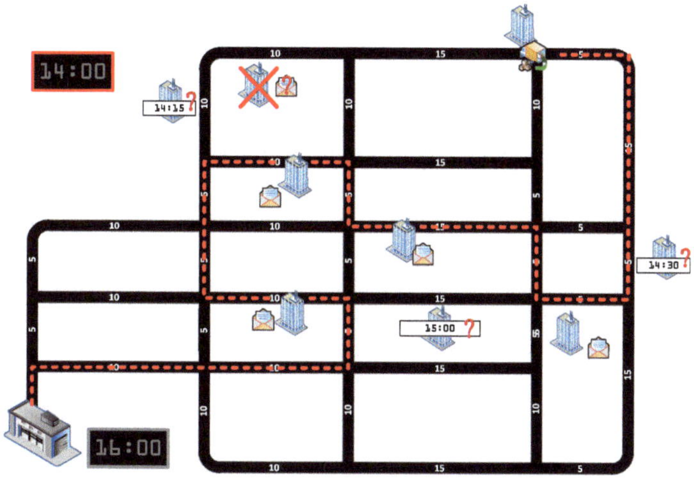

Fig. 8.4 Motivational example: plan 3

center of the service area. Even though the vehicle passes the request on the right, a confirmation is not possible because the request occurs after the vehicle already passed the location. Here, waiting at the current location would have had allowed to confirm this requests. As a result, not only the locations of the requests are important to anticipate, but also the request times and the vehicle's movements through the network. We call this the spatial and temporal *coverage* of the service area. This statement is reinforced by considering the third plan in Fig. 8.4.

To confirm the upper left customer request, the entire planned tour has to be changed. The confirmed request is the next customer to visit. As a result, the adjacent customer requesting at 2:15 p.m. can be included without additional travel time. Further, the vehicle reaches the center of the service area later and is therefore able to spend the remaining free 10 min of the time budget to include the central customer requesting at 3 p.m. without additional travel time. The conclusions of the example are as follows. Even though plan 3 has less free time budget to include new requests than plan 1, the coverage caused by the extended tour allows more confirmations. Further, the coverage also depends on the locations the vehicle passes at a certain time. The tours of plan 2 and plan 3 have the same duration. Nevertheless, for this example, plan 3 allows the confirmations of two additional future requests compared to plan 2. As a result, the free time budget, the spatial and temporal coverage of the service area indicate the expected number of future confirmations, i.e., the value of a post-decision state. We utilize this dependency to select and derive the anticipatory solution methods.

8.3 Outline

The second part of the book is outlined as follows. In Chap. 9, we formally define the dynamic routing problem and model it as a Markov decision process and recall and discuss the literature on vehicle routing problems with stochastic customer requests. We especially focus on the anticipation of area coverage in the solution approaches. In Chap. 10, we present the anticipatory solution approaches based on the literature review in Sect. 9.3. We especially focus on reactive anticipation. Therefore, we define additional reactive anticipatory approaches. In Chap. 11, we conduct an extensive computational evaluation for a variety of real-world sized instances differing in customer distribution, service area size, and ratio of dynamic customers. We depict the advantages of reactive anticipation and analyze the offline and online reactive anticipatory approaches in detail.

Chapter 9
SDVRP with Stochastic Requests

In the following, we rigorously model the presented routing application following the definition of Ulmer et al. (2017b). To describe dynamic decision making, we use a Markov decision process. For a mathematical formulation of the ex post mixed integer program, the reader is referred to Ulmer et al. (2015a). We finally give an overview on the regarding literature. We focus on the approaches' spatial and temporal anticipation.

9.1 Problem Statement

A vehicle serves customer requests in a service area. It starts its tour at a depot and has to return to the depot within the work shift. In the beginning, a set of early request customers (ERC) is given. These customers have to be served. During the shift, previously unknown late request customers (LRC) request service. Decision points occur by arriving at a customer. The dispatcher has to select the subset of new customer requests to be confirmed and the next customer to be visited. Besides traveling, waiting at customer locations is permitted. Confirmations and rejections are permanent. The shift ends when all ERC and confirmed LRC are served and the vehicle has returned to the depot. Within each problem realization, the dispatcher aims on maximizing the number of served LRC.

The notation for the problem is depicted in the upper part of Table 9.1. In the lower part of Table 9.1, the terminology required for the Markov decision process is defined. Let $T = \left[0 = t^{\text{init}}, 1, \ldots, t^{\text{limit}}\right]$ be the time horizon with initial time $t^{\text{init}} = 0$ and time limit t^{limit}. Let $\mathscr{A}$ be the service area. The vehicle starts and ends its tour in a depot $D \in \mathscr{A}$. The vehicle travels with a constant speed v. In the following, we describe a stochastic problem realization $\omega \in \Omega$ of the overall

© Springer International Publishing AG 2017

M.W. Ulmer, *Approximate Dynamic Programming for Dynamic Vehicle Routing*,
Operations Research/Computer Science Interfaces Series 61,
DOI 10.1007/978-3-319-55511-9_9

set of realizations Ω. Notably, in the decision making process, the customers are unknown before their request time. A problem realization $\omega \in \Omega$ consists of a set of customers $\mathscr{C}^\omega = \{C_1^\omega, \ldots, C_h^\omega\}$, according request times $t_i \in T$, and locations $l_i^{C^\omega} \in \mathscr{A}, \forall C_i^\omega \in \mathscr{C}^\omega$ as depicted in Eq. 9.1.

Table 9.1 Problem notations and MDP components

Parameter	Notation
Time horizon	$T = [0, \ldots, t^{\text{limit}}]$
Service area	$\mathscr{A}$
Depot	D
Vehicle speed	v
Overall set of realizations	Ω
Realization	$\omega \in \Omega$
Customers of realization ω	$\mathscr{C}^\omega = \{C_1^\omega, \ldots, C_h^\omega\}$
Early request customers	$\mathscr{C}_0^\omega$
Late request customers of a realization	$\mathscr{C}_+^\omega$
Spatial-temporal probability distribution	Ξ
Request time of customer $C_i^\omega \in \mathscr{C}^\omega$	$t_i \in T$
Location of customer $C^\omega \in \mathscr{C}^\omega$	$l_i^{C^\omega} \in \mathscr{A}$
Travel time between customers $C_1^\omega, C_2^\omega \in \mathscr{C}^\omega$	$\tau(l_1^{C^\omega}, l_2^{C^\omega})$
Decision points	$k = 0, \ldots, K-1$
Initial state	S_0
Termination state	S_K
State in decision point k	S_k
Decisions in decision point k	$X(S_k)$
Post-decision state in decision point k	S_k^x
Point of time in decision point k	$t(k)$
Vehicle position in decision point k	P_k
Customers to serve in decision point k	$\mathscr{C}_k$
Customer requests in decision point k	$\mathscr{C}_k^r = \omega_{k+1} \subset \omega$
Confirmed requests in decision point k	$\mathscr{C}_k^c$
Next customer to visit	C_{next}^k
Planned tour in decision point k, given decision x	Θ_k^x
Tour duration Θ_k^x	$\bar{\tau}(\Theta_k^x)$

$$\omega = \{(C_1^\omega, t_1^\omega, l_1^{C^\omega}), (C_2^\omega, t_2^\omega, l_2^{C^\omega}), \ldots, (C_h^\omega, t_h^\omega, l_h^{C^\omega})\} \quad (9.1)$$

The customers are divided in two temporal classes as shown in Eq. 9.2: early request customers $\mathscr{C}_0^\omega = \{C_i^\omega \in \mathscr{C}^\omega : t_i^\omega = 0\}$ requesting service in $t = 0$ and LRC $\mathscr{C}_+^\omega = \{C_i^\omega \in \mathscr{C}^\omega : t_i^\omega > 0\}$ requesting service in $t > 0$. The customers $\mathscr{C}_0^\omega$ are known in the beginning and must be served.

$$\mathscr{C}^\omega = \mathscr{C}^\omega_0 \cup \mathscr{C}^\omega_+ = \underbrace{\{C^\omega_1, \ldots, C^\omega_m\}}_{\mathscr{C}^\omega_0} \cup \underbrace{\{C^\omega_{m+1}, \ldots, C^\omega_h\}}_{\mathscr{C}^\omega_+} \tag{9.2}$$

Each customer $C^\omega \in \mathscr{C}^\omega$ is assigned to a vector (t^ω, l^{C^ω}) consisting of a request time $t^\omega \in T$ and a location $l^{C^\omega} \in \mathscr{A}$. The assignment follows a spatial and temporal probability distribution $(t^\omega, l^{C^\omega}) \sim \Xi : T \times \mathscr{A} \to [0, 1]$. The travel time between two customers is defined by $\tau(l_1^{C^\omega}, l_2^{C^\omega})$. Within each realization, the dispatcher aims on maximizing the number of confirmed LRC. In unlikely cases, where the tour duration to serve all ERC already exceeds the time limit, all ERC are served and no confirmation of dynamic requests is allowed.

9.2 Markov Decision Process Formulation

For the given problem, a decision contains both a confirmation and a movement action. To illustrate the decision making process, we model the problem as a Markov decision process. The required terminology is depicted in the lower part of Table 9.1. In an MDP, in each decision point k, a state S_k of a finite state space $\mathscr{S} = \{S_0, \ldots, S_q\}$ and a subset of possible decisions $X(S_k) \subseteq X = \{x_1, \ldots, x_r\}$ depending on state S_k are given. Each decision $x \in X(S_k)$ in state $S_k \in \mathscr{S}$ generates an immediate reward $R : \mathscr{S} \times X(S_k) \to \mathbb{R}$. The outcome of each combination $(S_k, x) \in \mathscr{S} \times X(S_k)$ is known beforehand and is defined as the post-decision state $S_k^x \in \mathscr{P} = \mathscr{S} \times X$. $\mathscr{P}$ denotes the post-decision state space. Given a post-decision state, a transition leads to a subsequent state $S_{k+1} = (S_k^x, \omega_k)$. S_{k+1} is determined by realization $\omega_k : \mathscr{P} \to \mathscr{S}$ of the set of (stochastic) problem realizations $\omega_k \in \Omega$. For the given problem, the initial state S_0^ω is defined by the set of ERC $\mathscr{C}_0^\omega$ of the realization ω. A decision point occurs initially when the vehicle starts at the depot and whenever the vehicle is located at a customer. A state at decision point k consists of the point of time $t(k) \in T$, the vehicle's position $P_k \in \mathscr{A}$, and a set of customers to visit $\mathscr{C}_k^\omega = \mathscr{C}_0^\omega(k) \cup \mathscr{C}_+^\omega(k)$ containing the not yet served subset of ERC $\mathscr{C}_0^\omega(k) \subseteq \mathscr{C}_0^\omega$ and the not yet served subset of confirmed LRC $\mathscr{C}_+^\omega(k) \subseteq \mathscr{C}_+^\omega$. Additionally for $k > 0$, a set of requests $\omega_{k-1} = \mathscr{C}_r^\omega(k) \subseteq \mathscr{C}_+^\omega$ that occurred between the last decision point $k-1$ and decision point k is given. Decisions $X(S_k)$ contain the confirmation action selecting the subset $\mathscr{C}_c^\omega(k) \subseteq \mathscr{C}_r^\omega(k)$ to confirm. As a movement action, the next customer $C_{\text{next}}^k \in \mathscr{C}_0^\omega(k) \cup \mathscr{C}_+^\omega(k) \cup \mathscr{C}_c^\omega(k) \cup \{P_k\}$ to visit is selected. If $C_{\text{next}}^k = P_k$, the vehicle idles at its current location for one time step $\bar{t}$. The idle time can be extended to w time steps by repeatedly setting $C_{\text{next}}^{k+j} = P_{k+j}$ for $0 \leq j < w$. The immediate reward $R(S_k, x) = |\mathscr{C}_c^\omega(k)|$ of a decision x given state S_k is defined by the cardinality of the confirmed subset of requests.

A decision x is feasible if there exists at least one feasible tour $\Theta_k^x = (\theta_1, \ldots, \theta_o)$ starting at the vehicle's position $l_1^\theta = P_k$, traveling to the next customer $\theta_2 = C_{\text{next}}^k$, including all remaining confirmed LRC and ERC, and ending at the depot $(\theta_o = D)$. The tour duration Θ_k^x is defined as $\bar{\tau}(\Theta)$ in Eq. 9.3.

$$\bar{\tau}(\Theta_k^x) = \sum_{i=1}^{o-1} \tau(l_i^\theta, l_{i+1}^\theta) \tag{9.3}$$

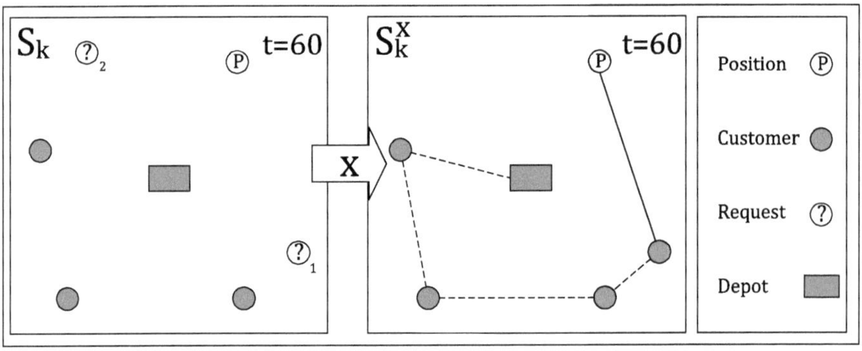

Fig. 9.1 State S_k before and post-decision state S_k^x after application of decision x

A planned tour Θ_k^x is feasible if $\bar{\tau}(\Theta_k^x) \leq t^{\text{limit}} - t(k)$, i.e., the tour duration of Θ_k^x allows returning to the depot within the time limit t^{limit}. A decision x results in a post-decision state S_k^x containing the time $t(k)$, the vehicle's location P_k, the next customer to visit C_{next}^k, and a set of remaining customers $\mathscr{C}_k^\omega \cup \mathscr{C}_c^\omega(k)$. The travel to the next customer and realization ω_{k+1} lead to a transition to decision point $k+1$. State S_{k+1} consists of position $P_{k+1} = C_{\text{next}}^k$, time $t(k+1) = t(k) + \tau(P_k, l^{C_{\text{next}}^k})$, and the set of customers $\mathscr{C}_{k+1}^\omega = \mathscr{C}_k^\omega \cup \mathscr{C}_c^\omega(k) \backslash \{C_{\text{next}}^k\}$. Further, ω_{k+1} provides a set of new requests $\omega_{k+1} = \mathscr{C}_r^\omega(k+1) = \{C_i^\omega \in \mathscr{C}_+^\omega : t(k) < t_i \leq t_{k+1}\}$. Because customer requests are stochastic, $\mathscr{C}_{k+1}^r$ is only revealed at decision point $k+1$ and unknown before. The overall realization $\omega \in \Omega$ is split regarding the time of the decision points as exemplarily depicted in Eq. 9.4.

$$\omega = \{\underbrace{(C_1, t_1, l_1^{C^\omega}), (C_2, t_2, l_2^{C^\omega})}_{\omega_0}, \underbrace{(C_3, t_3, l_3^{C^\omega})}_{\omega_1}, \ldots, \underbrace{(C_{h-1}, t_{h-1}, l_{h-1}^{C^\omega}), (C_h, t_h, l_h^{C^\omega})}_{\omega_{K-1}}\} \tag{9.4}$$

The MDP terminates in state S_K when all ERC and confirmed LRC are served and the vehicle has returned to the depot, i.e., $\mathscr{C}_K^\omega = \emptyset$, $t(K) \leq t^{\text{limit}}$, and $P_K = D$.

On the left side of Fig. 9.1 (Ulmer et al. 2015b), an exemplary state S_k is shown. In $t(k) = 60$, a set of three customers $\mathscr{C}_k^\omega = \{C_1, C_2, C_3\}$ has to be served. Two new requests $\mathscr{C}_r^\omega(k) = \{C_1^r, C_2^r\}$ are given. Decisions consist of four confirmation actions and up to six movement actions. In this exemplary case, the free time budget does not allow to confirm both requests. The applied decision x consists of confirmation action $\mathscr{C}_c^\omega(k) = \{C_1^r\}$, i.e., to accept request C_1^r and to reject request C_2^r. This leads to an immediate reward of $R_k(S_k, x) = 1$. Further, the next customer to visit C_{next}^k is set to C_1^r. The resulting post-decision state is depicted on the right side of Fig. 9.1.

S_k^x contains four customers, the next customer to visit is depicted by the bold line. The dashed line indicates a feasible tour Θ_k^x.

The dispatcher is aiming for a large number of confirmations, i.e., a high overall reward. Because the customer requests are stochastic, future rewards are not known beforehand. The objective is to find an optimal decision policy $\pi^* \in \Pi$ maximizing the expected number of confirmations.

Table 9.2 Literature classification

	Solution approach			Anticipation			
	Subset selection	Temporal	Spatial	Reactive	Explicit	Online	Offline
Psaraftis (1980)	n/a						
Bertsimas and Van Ryzin (1991)	n/a						
Tassiulas (1996)	n/a						
Gendreau et al. (1999)	n/a						
Swihart and Papastavrou (1999)	n/a						
Larsen et al. (2002)	n/a						
Mitrović-Minić and Laporte (2004)	n/a						
Thomas and White III (2004)	n/a	✓	✓	✓	✓	✓	
Bent and Van Hentenryck (2004)	n/a		✓		✓	✓	
Van Hemert and La Poutré (2004)			✓	✓	✓	✓	
Branke et al. (2005)		✓	✓	✓	✓		✓
Ichoua et al. (2006)	n/a		✓		✓		
Gendreau et al. (2006)	n/a						
Hvattum et al. (2006)	n/a		✓	✓	✓	✓	
Thomas (2007)		✓	✓		✓		
Ghiani et al. (2009)	n/a	✓	✓	✓	✓	✓	
Meisel (2011)	✓	✓	✓	✓	✓		✓
Ulmer et al. (2015a)	✓	✓			✓		
Ulmer et al. (2017b)	✓	✓	✓	✓	✓		✓
Ulmer et al. (2015b)	✓	✓	✓	✓	✓	✓	
Ulmer et al. (2017a)	✓	✓	✓	✓	✓	✓	✓

9.3 Literature Review

In the sequel, we recall SDVRPs presented in Sect. 7.5 considering stochastic customer requests. For these problems, decisions consider both routing and request confirmations. In the following, we describe the applied algorithms focusing on temporal and spatial consideration of area coverage and future requests. We highlight the publications introducing the selected approaches presented in Chap. 10.

Table 9.2 shows an overview of SDVRPs with stochastic customer requests (Ulmer et al. 2017a). We classify work regarding the solution approach and recall the degree of anticipation. A part of the work aims on maximizing the number of served customers given a time limit. Some problems require to serve all customers minimizing the travel time, waiting time, number of vehicles, or the deviation from time windows. For these problems, rejections are not possible, indicated by "n/a" in the subset selection column. For the other problems, we depict if an explicit subset selection is applied meaning that feasible requests may be rejected. This is achieved by the algorithms of Meisel et al. (2009, 2011), Ulmer et al. (2015a, b, 2017b).

As motivated in Chap. 8, anticipation of the spatial and temporal developments is desired. We recall the degree of anticipation of the applied solution approaches on the right side of Table 9.2 and classify them regarding their *spatial* and *temporal* consideration of area coverage and future requests on the left side of Table 9.2.

The initial work on stochastic customer requests is aiming on applying plain routing policies to reduce the expected travel times. First come, first serve-policies are applied by Psaraftis (1980); Bertsimas and Van Ryzin (1991); Swihart and Papastavrou (1999), and Larsen et al. (2002). Tassiulas (1996) partitions the service region and subsequently serves the subareas. Gendreau et al. (1999, 2006) combine tabu search and an adaptive memory with a rolling horizon algorithm to dispatch customer requests to a fleet of vehicles. Other approaches are straightforward waiting strategies (e.g., *wait at start*, *wait at end*), for instance applied by Mitrović-Minić and Laporte (2004). These approaches achieve implicit non-reactive anticipation, but do not explicitly consider future requests in decision making (compare Sect. 7.5) or spatial and temporal coverage of the service area.

Explicit non-reactive anticipation is mainly achieved by sampling. Within the execution phase, the approaches simulate a set of future events to evaluate current decisions. Sampling allows a more detailed consideration of future events, but requires significant computational effort within the execution phase. To anticipate stochastic customer requests in vehicle routing, future customer requests are sampled according to the spatial distribution or the graph. These requests are used to evaluate different routes and decisions. Bent and Van Hentenryck (2003); Bent and Van Hentenryck (2004) introduce a multiple scenario approach where customer requests are sampled and integrated in plans containing a set of routes. They only consider the spatial coverage of the routes neglecting temporal developments. This approach is also used by Flatberg et al. (2007) and Sungur et al. (2010). Hvattum et al. (2006) apply the MSA to real-world case study by using historical data of customer requests to minimize the expected travel time. Again, MSA only considers the spatial coverage of the service area. Notably, most of the sampling approaches decide about subset selection only implicitly. Explicit subset selection would lead to an exponential increase of the already high calculation times and even to computational intractability (Powell and Ryzhov 2012, p. 203ff). Ulmer et al. (2015a) propose a cost benefit heuristic (CBH) comparing the relative insertion time of a subset candidate with the relative gain of confirming a subset. If the ratio exceeds a threshold, the candidate subset is rejected. The threshold is determined via offline sampling and simulation. The area coverage is considered only implicitly by Ulmer et al. (2015a).

Beside sampling for non-reactive explicit anticipation, Thomas and White III (2004), Ichoua et al. (2006), Thomas and White III (2007), and Thomas (2007) propose waiting policies including information about future requests to achieve explicit consideration of temporal and spatial coverage. These policies select the routing, waiting locations, and waiting times according to potential future customers. Thomas (2007) introduces an anticipatory insertion waiting approach (AI). AI considers the center of gravity of the potential and feasible future requests in the selection of a suitable waiting location. Ghiani et al. (2012) compared AI with MSA by Bent and Van Hentenryck (2004) and achieved similar results, even though AI requires significantly less computational effort.

Explicit reactive anticipatory approaches can be divided in offline and online approaches. Offline approaches do not require extensive computational effort within the execution of the algorithm, but achieve a decision policy within a learning phase. For the given problem, offline algorithms are applied by Meisel et al. (2009, 2011). They use AVI and MVF to evaluate post-decision states achieving spatial-temporal consideration of the coverage. Offline reactive algorithms in vehicle routing generally suffer from the post-decision state space dimensionality because the values of post-decision states have to be stored for the execution phase (Powell 2011). As a result, Meisel et al. (2009, 2011) are only able to apply the approach for a graph containing 49 possible customer locations. Ulmer et al. (2017b) aggregate and partition the PDS space to apply AVI for a problem with arbitrary customer locations in the service area. With the anticipatory time budgeting approach (ATB), Ulmer et al. (2017b) introduce the first offline reactive approach explicitly considering both subset selection and (temporal) area coverage for problem settings of real-world size. ATB bases on temporal attributes and therefore not able to achieve explicit spatial anticipation.

Approaches providing online reactive anticipation are developed by Van Hemert and La Poutré (2004), Branke et al. (2005), and Ulmer et al. (2015b). Van Hemert and La Poutré (2004) evaluate PDSs by the number of regions the vehicle is still able to visit achieving spatial consideration of the coverage. Branke et al. (2005) use sampling to calculate the confirmation possibility of including a single request. They achieve spatial and temporal anticipation. For spatial anticipations, Ulmer et al. (2017a) samples ad hoc a set of request evaluating PDSs by the number of sampled requests feasible to be included immediately. Ulmer et al. (2015b) apply a rollout algorithm (RA) using sampled realizations to calculate the rewards to go. The RA enables spatial and temporal anticipation. Ulmer et al. (2017a) introduce a hybrid RA as a combination of offline ATB by Ulmer et al. (2017b) and online RA by Ulmer et al. (2015b).

For the presented SDVRP, we apply the approaches by Mitrović-Minić and Laporte (2004), Thomas (2007), Ulmer et al. (2015a), Ulmer et al. (2017b), Ulmer et al. (2015b), and Ulmer et al. (2017a). We recall the approaches in detail in Chap. 10. We additionally present reactive anticipatory derivations of approaches by Bent and Van Hentenryck (2004) and Van Hemert and La Poutré (2004).

Chapter 10
Solution Algorithms

In this chapter, we define the approaches for the SDVRP. We present an approach for (nearly) every class of anticipation. For the given problem, we were not able to construct a reactive and implicit anticipatory approach. The complexity of the problem and the instances did not allow to detect suitable attributes, parameters, and weights to evaluate a state without simulation. We treat the non-reactive approaches as benchmarks and focus on the reactive approaches. We show the required steps to achieve reactive anticipation for the given problem and depict the advantages and disadvantages of offline and online anticipation as presented in Sect. 6.6. The applied solution algorithms are depicted in the following.

- To evaluate the approaches regarding anticipation, we implement a *myopic* policy in Sect. 10.2.
- As non-reactive implicit approaches, we implement basic waiting policies in Sect. 10.3 (Mitrović-Minić and Laporte 2004).
- We present two non-reactive explicit approaches individually focusing on coverage and subset selection. We present anticipatory insertion (Ghiani et al. 2012) in Sect. 10.4 and the cost benefit heuristic (CBH, Ulmer et al. 2015a) in Sect. 10.5. AI waits at promising locations in the service area to maintain a high coverage and to efficiently include future requests. CBH rejects requests with insertion times above average. Therefore, CBH compares the reward of including a candidate subset of requests with the required time consumption. If the ratio of the insertion time exceeds the ratio of the rewards, the candidate subset is rejected. We apply simulation for the required parametrization of CBH.
- For reactive anticipation, we present four explicit approaches based on approximate dynamic programming to estimate the rewards-to-go for a particular decision state. The approaches differ in their focus on spatial and temporal anticipation. In Sect. 10.6, we present the offline anticipatory time budgeting approach (Ulmer et al. 2017b) approximating the expected number of future confirmations for a

© Springer International Publishing AG 2017
M.W. Ulmer, *Approximate Dynamic Programming for Dynamic Vehicle Routing*,
Operations Research/Computer Science Interfaces Series 61,
DOI 10.1007/978-3-319-55511-9_10

certain state of the problem. To achieve suitable aggregation, we follow Ulmer et al. (2017b) depicting and exploiting the dependencies between point of time, free time budget, and expected number of future confirmations. Regarding the literature review in Sect. 9.3, ATB uses only temporal attributes to anticipate future events.

- In Sect. 10.7, we present Ad Hoc Sampling (AHS) by Ulmer et al. (2017a), a reactive online ADP-approach based on Van Hemert and La Poutré (2004) and Bent and Van Hentenryck (2004). AHS samples a set of requests in the service area and evaluates post-decision states regarding the number of sampled requests which can be included ad hoc. Hence, AHS uses only spatial attributes to anticipate future events.
- In Sect. 10.8, we present an online myopic-based post-decision RA (Ulmer et al. 2015b). The RA simulates future decision points and transitions applying a myopic base policy.
- We further present a hybrid ATB-based post-decision RA drawing on ATB as (reactive anticipatory) base policy as proposed by Ulmer et al. (2017a). Both RAs achieve spatial-temporal anticipation. The ATB-based RA allows a significantly stronger temporal anticipation compared to the myopic-based RA.
- The approaches differ in the selected subset of requests, waiting decisions and anticipation. For all approaches, we draw on identical sequencing actions to allow comparison. We apply cheapest insertion (CI), recalled in Sect. 10.1.

In the remainder of this chapter, we first describe the routing policy and then the solution algorithms in detail.

10.1 Routing and Sequencing Decisions

In decision making, for every candidate subset of requests, the routing has to be determined for two reasons. First, the movement action has to be determined. Second, each potential candidate subset of confirmed requests has to be checked for feasibility regarding the time limit. In this section, we motivate the use of CI for initializing the tour and the determination of customer sequences. We briefly discuss the impact of improvement algorithms.

10.1.1 Subset Selection

Every algorithm draws on the same subset candidate procedure. Given state S_k and $\mathscr{C}_k^r$ with $|\mathscr{C}_k^r| = m$, at most 2^m PDSs have to be considered. Because the number of PDSs is exponentially dependent on the number of request candidates, we first check every request subset $\mathscr{C}_i(k) \subseteq \mathscr{C}_k^r$ with $\mathscr{C}_i(k) = \{C_i\}$, i.e., each individual request for feasibility. If the resulting tour is not feasible, C_i results in infeasibility for every other subset containing C_i. Hence, C_i can be excluded and only half of the overall subsets has to be considered. This procedure can be applied successively for every request reducing the number of considered subsets significantly.

10.1.2 Cheapest Insertion

For the presented problem, the decision space is still high-dimensional. Decisions contain the subset to be confirmed and the next customer to be visited or waiting respectively. For state S_k with $|\mathscr{C}_k| = o$ and $|\mathscr{C}_k^r| = m$, the decision space size $|X(S_K)| \leq 2^m \cdot (m + o)$ is the product of the number of subset candidates and the number of possible next customers to visit including waiting. To reduce the decision space, we apply a decomposition $\mathfrak{D} : \{X_i \subset X\} \rightarrow \{X_j \subset X\}$. Movement actions are conducted by cheapest insertion (CI) for all approaches.

CI is often applied in dynamic vehicle routing (e.g., Mitrović-Minić and Laporte 2004; Thomas 2007; Ghiani et al. 2009; Sungur et al. 2010; Ghiani et al. 2012). CI has two main advantages. First, it reflects the routing applied in practice. Usually, the dispatcher plans a sequence of the ERC in the beginning of the day and subsequently adds new customers to the existing tour. Second, the application of CI is efficient and allows fast feasibility checks and decision making within the execution phase. Given a set of new requests $\mathscr{C}_k^r$, all approaches evaluate every potential candidate subset of requests regarding feasibility and rewards. This results in a large number of candidate tours. Here, an efficient routing algorithm is mandatory to allow dynamic decision making.

The tour is initialized and updated regarding the following procedure: In $t = 0$, CI starts with a pre-decision tour $\Theta_0 = (D, D)$ only consisting of the depot. In every decision point k, a pre-decision tour Θ_k and a number of candidate subsets $\mathscr{C}_i(k) \subseteq \mathscr{C}_k^r$, $i = 1, \ldots, 2^{|\mathscr{C}_k^r|}$ of new requests are given. An exemplary pre-decision tour is depicted on the left side of Fig. 10.1. Three customers are already given. $\mathscr{C}_i(k)$ contains two requests. CI subsequently selects the cheapest request of the subset regarding the insertion time and adds it at the cheapest insertion position in the tour as depicted in the center of Fig. 10.1. This results in a candidate post-decision tour as seen on the right side of Fig. 10.1.

For $t = 0$, all requests are confirmed, i.e., $\mathscr{C}_c^{\omega}(0) = \mathscr{C}_0$ leading to decision x and post-decision tour $\Theta_k^x = (\theta_0, \ldots, \theta_o)$. If the selected movement action of x is not waiting, the first customer in the tour $C_{\text{next}} = \theta_1$ is the next customer to be visited. After the travel to the next customer, the visited customer is removed from Θ_k^x. This results in the pre-decision tour $\Theta_{k+1} = (\theta_1, \ldots, \theta_o)$. If waiting is applied, the pre-decision tour Θ_{k+1} is identical to Θ_k^x. If no customers are left to serve, i.e., $\Theta_k = (\mathscr{P}_k, D)$, but free time budget is left, the vehicle idles at the current location for all approaches except for the myopic approach. If no time is left, the vehicle returns to the depot.

A candidate subset and the according decision are considered feasible if the resulting candidate post-decision tour does not exceed the time limit. Because the decision space is reduced, CI may reject candidate subsets which are feasible regarding a different routing approach, e.g., an optimal TSP solution.

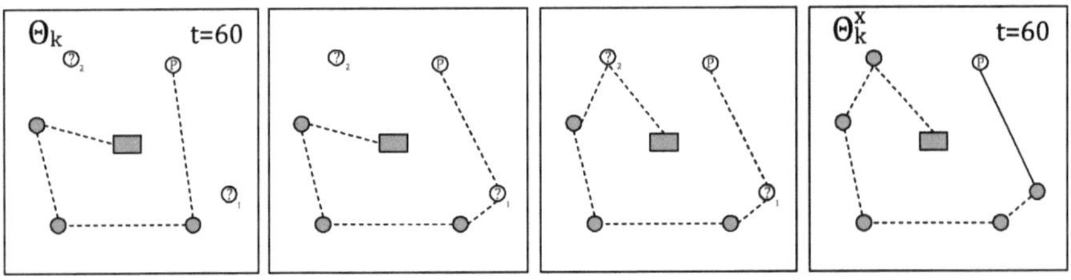

Fig. 10.1 Cheapest insertion

10.1.3 Improvements

In some work on SDVRPs, CI is additionally improved by metaheuristics. As an example, Ghiani et al. (2012) used a greedy randomized adaptive search procedure (GRASP, Feo and Resende 1995). We experimented with the GRASP proposed in Ghiani et al. (2012) and even minimized the tour duration in every post-decision state by solving the respective TSP. The intention of the improvement was to decrease the duration of the planned tour resulting in a larger amount of free time for future confirmations. Even though the required calculation time significantly increased, the achieved results in terms of confirmed requests only differed slightly compared to plain CI. In some cases, the number of confirmations even decreased. We interpret this behavior as a result from the difference in the nature of static and dynamic decision problems (Powell et al. 2000). As discussed in Sect. 2.5.5, an optimal static solution may lack in terms of flexibility. For the given problem, the reduction of the tour duration caused an early departure of regions in the service area. Subsequent requests in these regions could not be confirmed. Due to the results of these experiments, we decide to apply plain CI for all approaches.

10.2 Myopic Policy

In every decision point, the myopic policy selects the largest feasible subset of new requests (greedy). In cases, where several largest subsets are feasible, the myopic policy selects the subset with the smallest consumption of the time budget. As movement actions, the myopic policy neglects waiting and travels to the next customer in the tour. As a result, the myopic policy may lead to an early finish of the tour.

10.3 Non-reactive Implicit: Waiting Policies

For non-reactive implicit anticipation, we select waiting policies from Mitrović-Minić and Laporte (2004). We apply the *wait at start* (WAS) and *wait at end* (WAE) policy. WAS idles at the depot as long as free time budget is left. Therefore, the coverage increases over time and new requests can be efficiently inserted. WAE waits at the last customer before it returns to the depot. Both policies draw on myopic confirmations as described in Sect. 10.2.

10.4 Non-reactive Explicit: Anticipatory Insertion

Anticipatory Insertion (AI) was introduced by Ghiani et al. (2012) and is based on waiting strategies by Thomas (2007). For a dynamic routing problem with known customer locations, AI is able to achieve similar results as the sample-scenario approach by Bent and Van Hentenryck (2004). AI requires significantly less calculation effort in the execution phase and maintains the sequence of confirmed customers throughout the execution of the tour because of CI routing. The main idea of AI is to idle at certain locations in the service area to maintain a high coverage of the service area and therefore to insert new requests efficiently. AI draws on *myopic* confirmations.

To determine at which locations to wait, Ghiani et al. (2012) use the center-of-gravity (COG) longest wait strategy calculating the COG of all feasible potential future customers. The vehicle waits at the customer location which allows the latest departure time serving the remaining customers and a dummy customer located at the COG. The COG is recalculated in every decision point. For the presented problem, potential future customers are not known. To apply AI, in every decision point, we use sampling to determine the COG. We sample a set of m spatial realizations $F \subset \mathscr{A}$ of the respective distribution Ξ. Each sampled customer is checked for feasibility in the current tour. This results in a subset of feasible realizations $\bar{F} \subset F$. If $|\bar{F}| = 0$, the COG is set to the depot. Else, the COG is calculated regarding Eq. 10.1. Then, we proceed as described in Ghiani et al. (2012).

$$\mathrm{COG} = \left(\sum_{(a_x, a_y) \in \bar{F}} \frac{a_x}{|\bar{F}|}, \sum_{(a_x, a_y) \in \bar{F}} \frac{a_y}{|\bar{F}|} \right) \tag{10.1}$$

The procedure of AI is depicted in Fig. 10.2. Four customers have to be served. Six customers are sampled. Five of the sampled customers are individually feasible. The center of gravity lays within the center of these five customers. The customer determined as waiting location is depicted in black. As a result, the vehicle proceeds traveling to the next customer. At the next decision point a new waiting locations is determined. AI is able to anticipate spatial and temporal attributes about the potential customers and the routing.

Fig. 10.2 Anticipatory insertion: determination of the waiting location

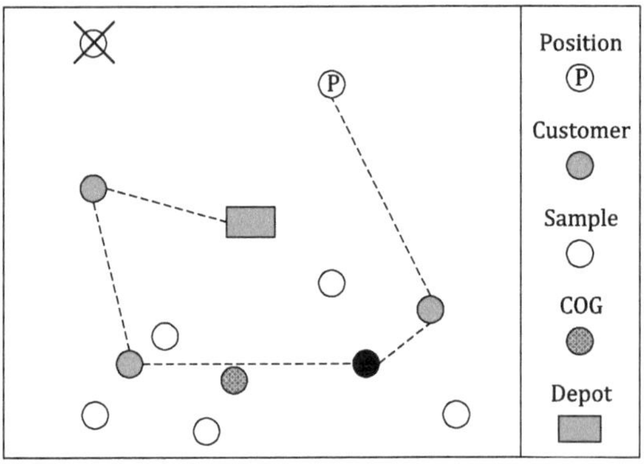

10.5 Non-reactive Explicit: Cost Benefit

With the cost benefit heuristic, Ulmer et al. (2015a) define an approach providing non-reactive explicit anticipation. This approach draws on expected future requests via sampling. CBH origins from the idea of Ichoua et al. (2000) weighting required travel time against resulting reward to decide about a diversion of the current tour. Ulmer et al. (2015a) transfer this idea to decide about the acceptance of candidate subsets. Subset candidates requiring a relatively high consumption of the free time budget are rejected. As a result, CBH conserves a high free time budget to include future customers, but on the expense of less immediate confirmations and a suffering of the service area coverage. Given a tour Θ_k and a candidate subset $\mathscr{C}_k^c$, CBH compares the insertion time ("cost") of decision x inserting $\mathscr{C}_k^c$ with the rewards ("benefit") as depicted in Eq. 10.2.

$$\kappa \cdot \frac{|\Theta_k^x|}{|\Theta_k|} \geq \frac{\bar{\tau}(\Theta_k^x)}{\bar{\tau}(\Theta_k) + d^*} \tag{10.2}$$

On the left side of Eq. 10.2, the relative benefit, i.e., the increase in customers to serve caused by the confirmed requests of a decision is calculated. On the right, the duration of the new tour compared to the current tour is calculated. Because the average insertion time of adding a customer is dependent on the instances, the parameters κ and d^* allow tuning regarding the instances, e.g., via simulation. If the tuning is done without consideration of the instances, CBH provides only implicit anticipation. If instance realizations are simulated to derive suitable parameter settings, CBH is explicitly anticipatory. Parameter κ scales the benefit compared to the costs. Parameter d^* defines a free amount of time budget relative to the tour length. This enables the insertion of customer requests especially in the beginning, when the tour might be short and insertion times are above-average. CBH indirectly considers the insertion time of requests and therefore the coverage. If the parameters are tuned by simulation, the temporal development of routing and requests is considered as well. Hence, CBH allows spatial and temporal anticipation.

10.6 Offline Reactive Explicit: ATB

In this section, we present anticipatory time budgeting by Ulmer et al. (2017b) proving offline reactive explicit anticipation. For problem instances of real-world size, the number of states, decisions, post-decision states, and transitions is vast. The expected future rewards cannot be calculated exactly, but can only be approximated. As described in Sect. 6.5, this approximation can be achieved by approximate dynamic programming (Powell 2011) and offline by approximate value iteration. ATB draws on the concepts of ADP and AVI for an offline policy derivation. As an offline approach, the calculation of the expected future confirmations or values is conducted in a learning phase and the achieved results can be efficiently applied within the execution phase. During the execution phase, ATB draws on the values allowing fast decision making. To store the values, an aggregation of the high-dimensional post-decision state space is necessary.

In Sect. 10.6.1, we recall the illustration by Ulmer et al. (2017b) that the expected number of future customers significantly depends on the point of time and the free time budget. We use these dependencies to motivate the aggregation of a post-decision state to a 2-dimensional vector containing point of time and free time budget. To evaluate the vectors, we use partitioning techniques described in Sect. 6.5. Such a partitioning defines a lookup table. A value is assigned to each entry of the LT. Since a direct calculation of the values for an entry is not possible, we estimate the values by approximate value iteration. Until now, we were only able to include spatial information in an offline reactive anticipatory approach. In Sect. 10.6.2, we briefly discuss the measures taken to examine the impacts of including additional spatial information.

10.6.1 Aggregation and Partitioning

Every value calculated in the learning phase has to be stored to access it in the execution phase. For the given problem, the number of post-decision states $S_k^x \in \mathcal{P}$ is large because customer requests can occur at any point in the service area. Further, in approximate value iteration, the more frequent a value is accessed and updated, the more accurate its approximation gets. If values are only accessed sparsely, the approximation and the solution quality might be impaired (Barto 1998, p. 193). Hence, the state space representation has to be of reasonable size to allow both efficient storage and effective approximation.

In practice, dispatchers merely consider single future requests but aim on budgeting the limited time efficiently. Decisions depend on the required amount of time to insert a customer (*insertion time*), the remaining *free time budget*, and the current *point of time*. A high free time budget combined with a low insertion time leads to a large number of expected confirmations. Nevertheless, insertion time and free time budget are dependent. Figure 10.3 shows the influencing factors on insertion time and free time budget and their dependencies for a certain point of time

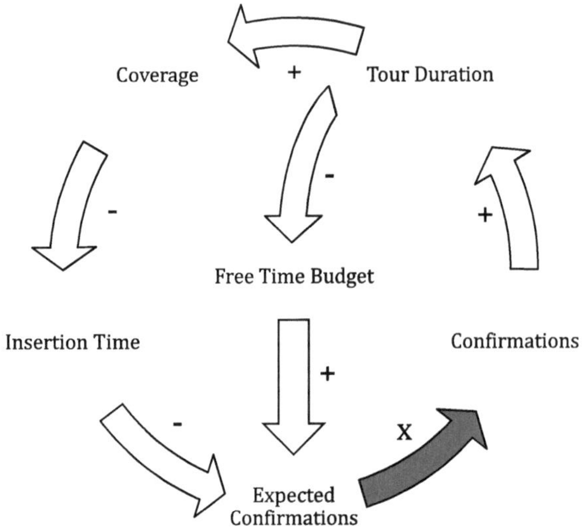

Fig. 10.3 Influencing factors on the expected number of future confirmations (Ulmer et al. 2017b)

(Ulmer et al. 2017b). Notably, the point of time itself is another substantial influencing factor for the expected number of future confirmations. The factors and the fortitude of the depicted dependencies in Fig. 10.3 are significantly influenced by the point of time.

The influence of a factor is indicated by an arrow. If a plus symbol is assigned to the arrow, an increase in the influencing factor results in an increase in the influenced factor. The minus symbol indicates an opposed effect. The gray arrow assigned with an x shows the point in the system where the dispatcher is able to control the dynamics. The dispatcher decides about the requests to be confirmed and the according routing considering the decision's impact on the expected number of future confirmations and the immediate confirmations. The applied decision then results in an adapted tour and a changed tour duration.

The insertion time significantly depends on the *coverage* of the service area (Branke et al. 2005). For the purpose of presentation, we assume coverage to be the percentage of the service area where customer requests can be inserted in the tour with a reasonable insertion time, i.e., an insertion time lower than a certain threshold. An increase in coverage results in a decrease of the insertion time. A decrease in insertion time leads to an increase in the expected number of future confirmations.

The coverage is dependent on the *tour duration* through the remaining early request and confirmed customers. If the tour duration increases, the coverage increases as well. As a result, a large number of immediate confirmations results in a long tour. The long tour allows low insertion times. This may increase the expected number of future confirmations. Nevertheless, a long tour duration results in a low amount of free time budget. Even though the insertion time is low because of the high coverage, the expected number of future confirmations decreases. Dispatchers experience a tradeoff between tour duration, i.e., current confirmations and free time budget or future confirmations respectively. To maximize the overall

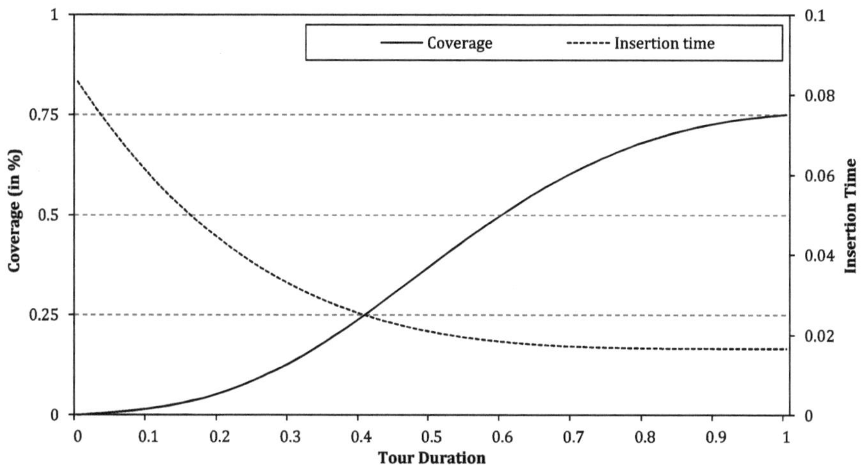

Fig. 10.4 Dependencies of coverage, tour duration, and insertion time

number of confirmations, dispatchers have to consider how a decision impacts the resulting insertion time and free time budget.

In Figs. 10.4, 10.5, 10.6 and 10.7, we show the dependencies on the expected number of future confirmations (Ulmer et al. 2017b). For the purpose of presentation, we depict the dependencies of coverage, tour duration, insertion time, point of time, free time budget, and expected number of future confirmations in an exemplary and idealized way. Evidently, the dependencies are influenced by a variety of factors and are especially dependent on the instance under consideration.

The dependencies between the tour duration, the coverage, and the resulting insertion time are depicted in Fig. 10.4. The tour duration $\bar{\tau}$ is depicted on the x-axis. $\bar{\tau} = 0$ represents a tour with zero travel time, $\bar{\tau} = 1$ a tour requiring the whole time budget to conduct. $\bar{\tau} = 1$ is only feasible in the beginning of the shift resulting in no free time budget. The coverage is depicted on the left y-axis. 0 represents no coverage at all, while 1 represents full coverage of the whole service area. We assume that it takes a certain duration of the tour to establish an initial coverage. Then, the coverage increases simultaneously with the extension of the tour. Given a long tour duration, the increase subsequently diminishes. A full coverage will not be achieved. The according insertion time relative to the overall time budget is shown on the right y-axis. It is dependent on the coverage. The first extension of the tour reduces the insertion time significantly. This impact decreases if the tour duration is already long. As a result of these dependencies, we can utilize the tour duration as an indicator for the insertion time and coverage.

Figure 10.5 shows the average behavior of tour duration and confirmations over time assuming a given confirmation and routing policy of the dispatcher. The (average) tour duration depend on the point of time t and is depicted by the solid line. Point of time $t = 0$ represents the beginning of the shift, $t = 1$ represents the end. The customer confirmations over time are shown on the right y-axis. The dashed line

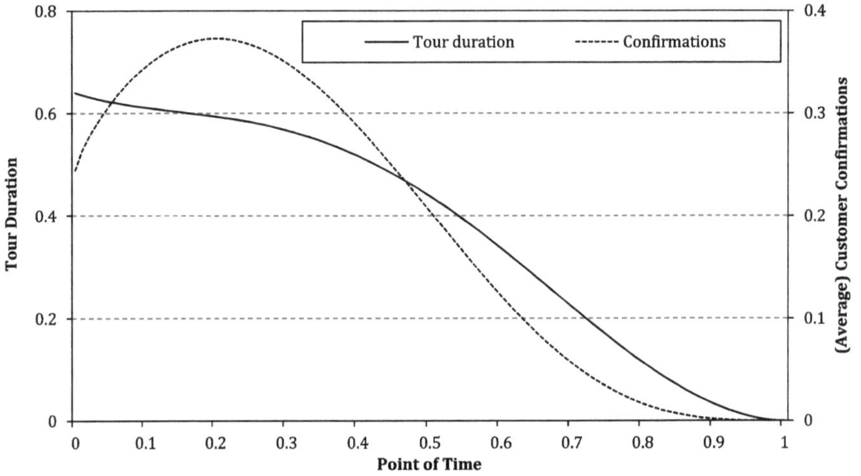

Fig. 10.5 Dependencies of point of time, confirmations, and tour duration (Ulmer et al. 2017b)

shows the (average) confirmations at a certain point of time. In this exemplary depiction, the number of confirmations decreases over time. In $t = 0$, a tour is already given because of initial early request customers. The duration decreases with the number of customers already served and increases with new confirmations. Since dynamic requests are confirmed, the duration of the tour is kept almost constant for some time. Later on, it will decrease nearly linearly. Towards the end of the shift, the vehicle serves the remaining customers and returns to the depot. For this example, we assume a given decision policy for confirmation and routing decisions. Nevertheless, decision making influences the tour duration significantly.

The extrema are shown in Fig. 10.6. The point of time is depicted on the x-axis. On the left y-axis, the tour duration is shown. The right y-axis represents the according free time budget b, calculated by the point of time and the tour duration: $b = 1 - t - \bar{\tau}$. The dashed line shows the minimal $\bar{\tau}$ over time. This minimum is provided by a policy rejecting all dynamic requests. The vehicle constantly serves the early request customers and $\bar{\tau}$ decreases linearly. The vehicle arrives early at the depot. The solid line represents the maximal possible $\bar{\tau}$ for every point of time. For the single points of time, this may be achieved by different policies. Evidently, $\bar{\tau} \leq 1 - t$. The minimal and maximal amount of free time budget are depicted accordingly. As we can see, a corridor of free time budget spans for every point of time t.

For $t = 0.5$, Fig. 10.7 shows this corridor on the x-axis. For this exemplary case, the maximal free time budget for $t = 0.5$ is $b = 0.34$ because the tour at least consists of the not yet served early request customers. The minimal free time budget is $b = 0$ meaning that all of the time budget is already consumed by previous confirmations. The expected insertion time for future requests, depicted on the left y-axis, directly depends on the free time budget resulting from the dependencies of free time budget and tour duration. Hence, the more free time budget is provided, the higher the

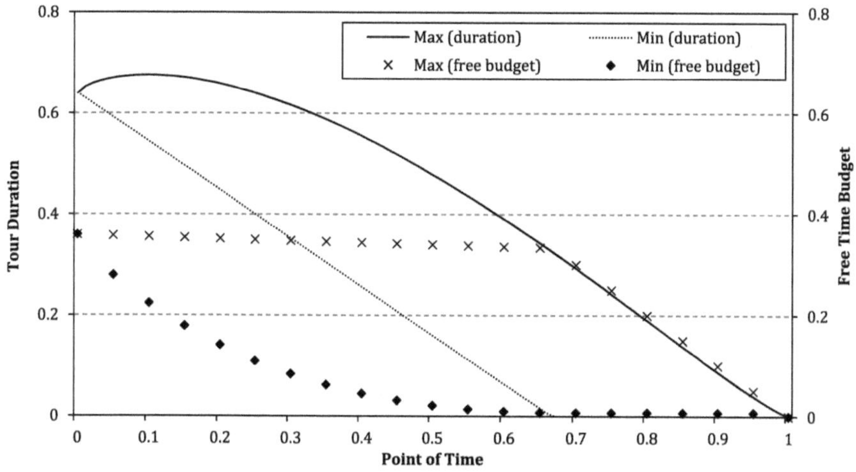

Fig. 10.6 Dependencies of point of time, tour duration, and free time budget (Ulmer et al. 2017b)

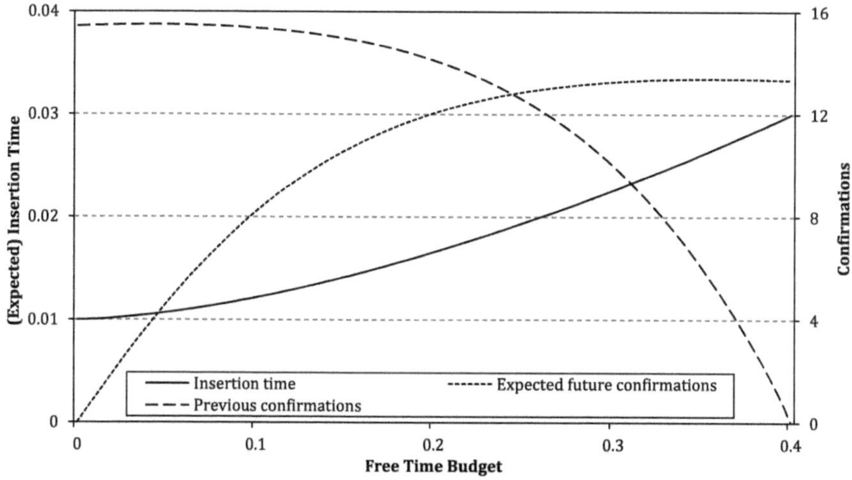

Fig. 10.7 Impact of insertion time, and free time budget on expected number of future confirmations (Ulmer et al. 2017b)

insertion time gets. Since the future insertion time depends on the applied policy, it may differ over the remaining points of time. Hence, the depiction in Fig. 10.7 is a simplification. The free time budget results from previous confirmations before $t = 0.5$. Given $b = 0$, many requests are included. Some of them may be expensive because they are not close to already existing customers and extend the tour duration significantly. The rejection of these requests leads to a substantial increase of the free time budget. For $b = 0.34$, all previous requests have been rejected. The expected number of future confirmations depends on the free time budget and the expected insertion time. Given $b = 0$, the insertion time is low, but no free time budget is remaining to insert a request. Given a high budget b, the insertion time increases significantly and the expected number of future confirmations stagnates. The dispatcher aims on maximizing the overall number of confirmations. If the expected number of future confirmations per point of time and free time budget is provided in every decision point, an optimal decision can be taken maximizing the sum of immediate confirmations and expected future confirmations (Bellman 1956).

As a result of the above considerations, the point of time and the free time budget can be used as indicators for the expected number of future confirmations. Nevertheless, the presented dependencies may be distorted by the instance settings. For example, a heterogeneous spatial request probability distribution in the service area impacts the dependency of tour duration and coverage. If requests are clustered in certain regions of the service area, the number of future confirmations may depend on the vehicle's and customers' locations. Further, the coverage, tour duration, and according insertion times may change unpredictably, especially when a vehicle leaves a cluster (Meisel 2011, p. 210f). Hence, an analytical calculation of the expected number of future confirmations seems not possible.

However, the expected number of future confirmations mainly depends on the point of time and the free time budget. The free time budget $b(k)$ is the difference of remaining time and tour duration as depicted in Eq. 10.3.

$$b(k) = t^{\text{limit}} - t(k) - \bar{\tau}(\Theta_k^x) \tag{10.3}$$

We use an aggregation $\mathfrak{A} : \mathscr{P} \rightarrow \mathscr{Q} \subsetneq \mathbb{N}^2$ to represent high-dimensional post-decision states $\mathfrak{A}(S_k^x) = p^k$ by 2-dimensional vectors $p^k = (t(k), b(k)) \in \mathscr{Q}$ only containing the numerical parameters point of time and free time budget. The resulting representation $\mathscr{Q}$ is defined in Eq. 10.4.

$$\mathscr{Q}(S_k) = (t(k), b(k)) \tag{10.4}$$

The aggregation results in a two-dimensional state space. We use the canonical Euclidean distance measure between the vectors. To partition the vector space, we apply partitionings $\mathscr{I}$ based on static lookup tables, weighted lookup tables, and dynamic lookup tables as described in Sect. 6.5.

10.6.2 *Extending the AVI-Vector Space*

To this point of time, an inclusion of spatial attributes in the (offline) AVI-vector space seems not possible. We have experimented with partitionings of the service area in regions as proposed by Schmid (2012) and we have included the vehicle's location and the number of customers per region in the vector space. We further have generated a set of customer representatives. To this end, a set of customers has

Algorithm 5: Ad Hoc Sampling

Input: State S_k, Decisions $x_1, \ldots, x_n$, Rewards $R(S_k, x_j)$, PDSs $\mathscr{P}_k = (S_k^{x1}, \ldots, S_k^{xm})$, PDS-Tours $\Theta(S_k^{x1}), \ldots, \Theta(S_k^{xm})$, Sample Sets $\hat{\mathscr{C}}_1, \ldots, \hat{\mathscr{C}}_n$, Time Limit t^{limit}
Output: Decision x^*

for all $S^x \in \mathscr{P}_k$ **do**
  $i \leftarrow 1$
  $\hat{V}(S^x) \leftarrow 0$
  `// Evaluation`
  while $(i \leq m)$ **do**
    $\Theta \leftarrow \Theta(S_k^x)$
    while $\hat{\mathscr{C}}_i \neq \emptyset$ **do**
      if $\bar{\tau}(CI(\Theta, \hat{\mathscr{C}}_i.getFirst())) + t(k) \leq t^{limit}$ **then**
        $\Theta \leftarrow CI(\Theta, \hat{\mathscr{C}}_i.getFirst())$
        $\hat{V}(S_k^x) \leftarrow \hat{V}(S_k^x) + \frac{1}{m}$
      end
      $\hat{\mathscr{C}}_i \leftarrow \hat{\mathscr{C}}_i \backslash \{\hat{\mathscr{C}}_i.getFirst()\}$
    end
    $i \leftarrow i + 1$
  end
  `// Selection`
  $R^* \leftarrow 0$
  for all $S_k^x \in \mathscr{P}^i$ **do**
    if $(R(S_k, x) + \hat{V}(S_k^x) \geq R^*$ **then** $x^* \leftarrow x$
  end
end
return x^*

been sampled regarding Ξ and "typical" customer locations have been determined by a k-medoid clustering algorithm (Vinod 1969). We have included the information about the representatives in the aggregation. We have further used fuzzy sets (Zadeh 1965) to evaluate similar customer sets equally. All spatial consideration led on the one hand to a tremendous increase of the vector space. On the other hand, an improvement of the solution quality could not be achieved. We conclude that (global) dependencies of spatial information and a PDS's values are difficult to derive. As a result, offline spatial anticipation is challenging.

10.7 Online Reactive Explicit: Ad Hoc Sampling

ATB allows anticipation from a global perspective only drawing on temporal attributes. To allow a local view of particular areas of the MDP and to consider spatial information, we present the online reactive ad hoc sampling approach by Ulmer et al. (2017a). AHS is a VFA enabling to evaluate PDSs regarding the locations of the vehicle, the customers, and the planned tour. AHS is inspired by

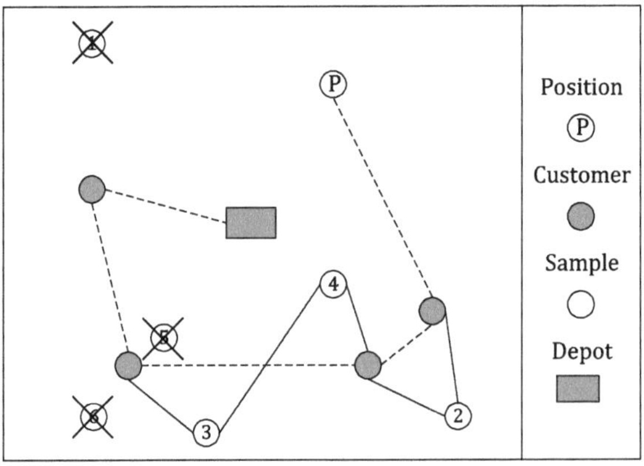

Fig. 10.8 Ad Hoc sampling

Bent and Van Hentenryck (2004) and Van Hemert and La Poutré (2004). Bent and Van Hentenryck (2004) use a set of sampled customers to evaluate current plans. Van Hemert and La Poutré (2004) evaluate plans regarding the number of "fruitful regions" with high request probabilities the vehicle is still able to visit given the free time budget. Ulmer et al. (2017a) combine the two approaches to AHS. The main idea of AHS is to evaluate the spatial coverage of a PDS by sampling a set of requests and examining the number of requests which can be inserted ad hoc.

The procedure of AHS is described in Algorithm 5. Let state S_k, decisions $x_1, \dots, x_n$, rewards $R(S_k, x_j)$, PDSs $\mathscr{P}_k = (S_k^{x_1}, \dots, S_k^{x_m})$, resulting tours $\Theta(S_k^{x_1})$, $\dots$, $\Theta(S_k^{x_m})$, ordered sets of sampled customers $\hat{\mathscr{C}}_1, \dots, \hat{\mathscr{C}}_n$, and the time limit t^{limit} be given. To evaluate a PDS, for every sample set $\hat{\mathscr{C}}_i$ the number of feasible sampled customers is determined. Therefore, the customers of the ordered set are subsequently inserted in the tour Θ via CI and then removed from $\hat{\mathscr{C}}_i$. Command $\hat{\mathscr{C}}_i$.*getFirst*() provides the first customer of set $\hat{\mathscr{C}}_i$. If an insertion is feasible, Θ is updated and the value $\hat{V}(S_k^x)$ is incremented by $\frac{1}{m}$. If a sampled customer cannot be inserted feasibly, the customer is skipped and the value remains. As a result, the value of a PDS is the average number of included sampled customers over all sets. The decision is selected that maximizes immediate reward $R(S_k, x)$ and approximate value $\hat{V}(S_k^x)$.

An exemplary evaluation for a PDS is shown in Fig. 10.8. For the purpose of presentation, we assume $m = 1$. The set of sampled customers $\hat{\mathscr{C}}_1 = (\hat{C}_1, \hat{C}_2, \dots, \hat{C}_6)$ contains six customers. AHS now subsequently checks the sampled customers for feasibility. The insertion of customer $\hat{C}_1$ violates the time limit. Hence, $\hat{C}_1$ is skipped and crossed out. Customers $\hat{C}_2, \hat{C}_3, \hat{C}_4$ can be inserted incrementing the value $\hat{V}(S_k^x) = 3$. $\hat{C}_5, \hat{C}_6$ are not feasible to include and are crossed out as well. Notably, AHS does not analyze every possible subset of sampled requests but subsequently inserts the requests in the order of $\hat{\mathscr{C}}_1$. Because AHS is an online approach and the number of sampled ad hoc requests is usually high, an explicit subset consideration of every sample set is computationally intractable.

10.8 Online/Hybrid Reactive Explicit: Rollout Algorithm

The two presented reactive anticipatory approaches ATB and AHS both only consider a single aspect of the spatial-temporal attributes and developments of a PDS. ATB allows the integration of future transitions and decisions but merely represents the temporal attributes point of time and free time budget. Customer locations and the planned tour are not considered explicitly. AHS allows an explicit consideration of customer locations and the planned tour but only for the current point of time. A consideration of the development over time is not given. To allow spatial and temporal anticipation, we apply a post-decision RA as defined in Sect. 6.3. The post-decision RA simulates future transitions and decisions by conducting the planned tour within the simulation. Hereby, the RA maintains the high level of detail regarding the customer locations and the tour.

Figure 10.9 shows (a part of) an exemplary simulation within RA for the presented problem. Given a PDS to evaluate shown on the left, RA simulates the "rest of the shift". In Fig. 10.9, the next three decision points within the simulation are depicted as PDSs $\hat{S}^x_{k+i}$, $i = 1, 2, 3$. The gray requests indicate the sampled requests within the simulation. In the first decision point sampled, the vehicle reached the next location and two new sampled request occur. The RA selects one of the requests to be included. Then, the RA simulates the travel to the next customer, one of the sampled requests. Again, the simulation provides two new sampled requests and the RA includes one of them. This procedure repeats until the terminal state is reached, i.e., all customers and confirmed sampled requests are served and the vehicle has returned to the depot. Notably, this is not depicted in Fig. 10.9. If we assume that after PDS $\hat{S}^x_{k+3}$ no further request is confirmed, the value of the PDS S^x_k for this sample run is $\hat{V}(S^x_k) = 2$.

As seen in the example simulation in Fig. 10.9, decisions about the subset of requests to select and routing have to be made within every simulation run. For routing, CI is applied as described in Sect. 10.1. For subset selection, we utilize two different base policies.

10.8.1 Myopic-Based Rollout Algorithm

The first base policy we apply is the myopic policy described in Sect. 10.2 resulting in RA X^m as proposed by Ulmer et al. (2015b). This policy always selects the largest feasible subset of sampled requests. Even though the base policy is myopic, the RA X^m itself is anticipatory drawing on the base policy for evaluation of the PDSs.

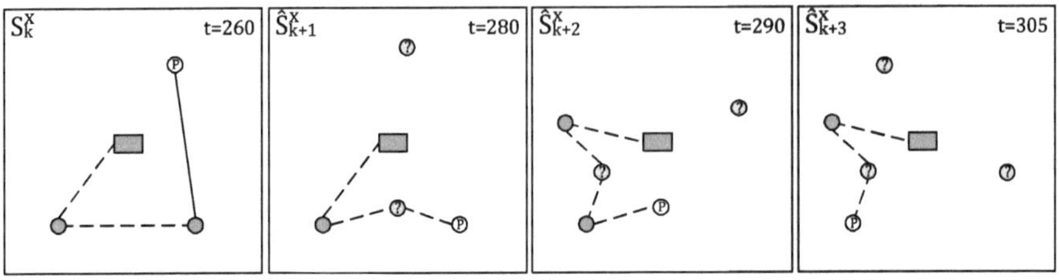

Fig. 10.9 Exemplary rollout-simulation

10.8.2 ATB-Based Rollout Algorithm

As shown in the Example in Sect. 6.6.3, the base policy has a significant impact on the achieved policy. An already anticipatory base policy strengthens the approximation of the RA. For the given problem, the number of decision points and PDSs in the RA is vast. To avoid computational intractability, the base policy has to enable an efficient application within the simulation. Hence, for the hybrid RA, we select the offline reactive ATB as base policy resulting in RA X^a. Considering the SDP solved within the RA, the use of ATB extends the number of considered transitions and decisions significantly. The combination of offline and online ADP-methods enables a general, global consideration of the SDP as well as a detailed, local consideration (Ulmer et al. 2017a).

Chapter 11
Computational Evaluation

In this chapter, we evaluate and analyze the approaches. We first define the set of test instances in Sect. 11.1 and tune the algorithm parameters in Sect. 11.2. In Sect. 11.3, we compare the approaches providing non-reactive anticipation with the offline reactive anticipatory ATB. We use the findings to analyze ATB in detail in Sect. 11.4 and to compare ATB with the online reactive anticipatory approaches in Sect. 11.5.

11.1 Instances

The instances are defined by Ulmer et al. (2017b). The generation of the instances and the according parameters are described in the following. The closed service area $\mathscr{A} \subsetneq \mathbb{R}^2$ is rectangular defined by the lower left point $(0, 0)$ and the upper right point $(x_{\max}, y_{\max}) \in \mathbb{R}^2$. Time is represented minute by minute, $T = \{0, 1, \ldots, t^{\text{limit}}\}$. The expected number of overall customers is $n = \mathbb{E}_{\omega \in \Omega} |\mathscr{C}^{\omega}|$. The expected number of early request customers $\mathbb{E}|\mathscr{C}^0| = n_0$ depends on the degree of dynamism $dod \in [0, 1]$ as depicted in Eq. 11.1 (Larsen et al. 2002).

$$n_0 = (1 - dod) \cdot n \tag{11.1}$$

The number of customers and the request times for a realization are generated by a Poisson process $\mathfrak{P}$ (Haight 1967). The number of ERC is generated by $\mathfrak{P}(n_0)$. The probability distribution Ξ for request times and locations is divided into two independent probability distributions. Request times of late request customers are (discretely) uniformly distributed over time $t \sim U_{\mathbb{Z}}[1, t^{\text{limit}} - 1]$. Customer locations $l^{C^{\omega}} \in \mathscr{A}$ are realizations $l^C \sim \mathscr{F}$ of the spatial probability distribution $\mathscr{F} : \mathscr{A} \to [0, 1]$.

© Springer International Publishing AG 2017

M.W. Ulmer, *Approximate Dynamic Programming for Dynamic Vehicle Routing*, Operations Research/Computer Science Interfaces Series 61,
DOI 10.1007/978-3-319-55511-9_11

A realization of the request time is again conducted by a Poisson process $\mathfrak{P}$ for every minute $0 < t < t^{\text{limit}}$. Given two points of time $0 < t^j < t^h < t^{\text{limit}}$, this results in an expected number of customers of $n_{t^j}^{t^h} = \mathbb{E}_{\omega \in \Omega} \left|\{C_i^\omega \in \mathscr{C}_+^\omega : t^j < t_i \leq t^h\}\right|$ requesting in time $t^j < t_i \leq t^h$ as described in Eq. 11.2.

Table 11.1 Instance parameters

Parameter	Setting
Time limit	$t^{\text{limit}} = 360\,\text{min}$
Expected number of customers	$n = 100$
Vehicle speed	$v = 25\,\text{km/h}$
Service area	$\mathscr{A}_{15}$, $\mathscr{A}_{20}$
Degree of dynamism	$dod = 0.25, 0.50, 0.75$
Spatial distribution	$\mathscr{F}_U$, $\mathscr{F}_{2C}$, $\mathscr{F}_{3C}$

$$n_{t^j}^{t^h} = dod \cdot n \cdot \frac{t^h - t^j}{|T| - 2} \tag{11.2}$$

The travel time $\tau : \mathscr{A} \times \mathscr{A} \rightarrow \mathbb{N}$ for two customers $C_1^\omega, C_2^\omega \in \mathscr{C}^\omega$ with locations $l_1^{C^\omega} = (a_1^x, a_1^y), l_2^{C^\omega} = (a_2^x, a_2^y) \in \mathscr{A}$ is Euclidean and rounded up to minutes as depicted in Eq. 11.3. Parameter v represents the vehicle's speed. The minimal travel time is set to $\bar{t} = 1\,\text{min}$.

$$\tau(l_1^{C^\omega}, l_2^{C^\omega}) = \max\left(\left\lceil \frac{((a_1^x - a_2^x)^2 + (a_1^y - a_2^y)^2)^{1/2}}{v} \right\rceil, 1\right) \tag{11.3}$$

The quantities of the instances are derived from Bent and Van Hentenryck (2004); Hvattum et al. (2006); Thomas (2007), and Meisel (2011). The instance parameters are listed in Table 11.1. The time limit is set to $t_{max} = 360\,\text{min}$. We test the approaches for a large ($\mathscr{A}_{20} : x_{\max} = 20\,\text{km}$, $y_{\max} = 20\,\text{km}$) and small service area ($\mathscr{A}_{15} : x_{\max} = 15\,\text{km}$, $y_{\max} = 15\,\text{km}$). The vehicle travels with a speed of $v = 25\,\text{km/h}$. The depot is located in the center of the area $D_{20} = (10, 10)$, respectively, $D_{15} = (7.5, 7.5)$. The average number of customer requests per day is $n = 100$. We examine instances with a small ($dod = 0.25$), moderate ($dod = 0.50$), and large ($dod = 0.75$) number of dynamic requests. We define three spatial distributions. We consider uniformly distributed customers ($\mathscr{F}_U$) and customers grouped in two ($\mathscr{F}_{2C}$) or three clusters ($\mathscr{F}_{3C}$). Within the clusters, the customers are normally distributed. Given $\mathscr{A}_{20}$, an exemplary customer setting with 50 realized customers for $\mathscr{F}_{2C}$ and $\mathscr{F}_{3C}$ is shown in Fig. 11.1. $\mathscr{F}_{2C}$ and $\mathscr{F}_{3C}$ are defined to analyze the impact of a heterogeneous spatial distribution to decision making, keeping in mind that ATB only draws on temporal attributes and neglects explicit spatial information and AHS only allows spatial anticipation without consideration of temporal developments. We assume that ATB is able to perform well for $\mathscr{F}_U$ and for the symmetrical $\mathscr{F}_{2C}$ because the spatial information about the sequence of cluster visits does not influence the temporal attributes and the regarding outcomes. For $\mathscr{F}_{3C}$, the sequence significantly impacts

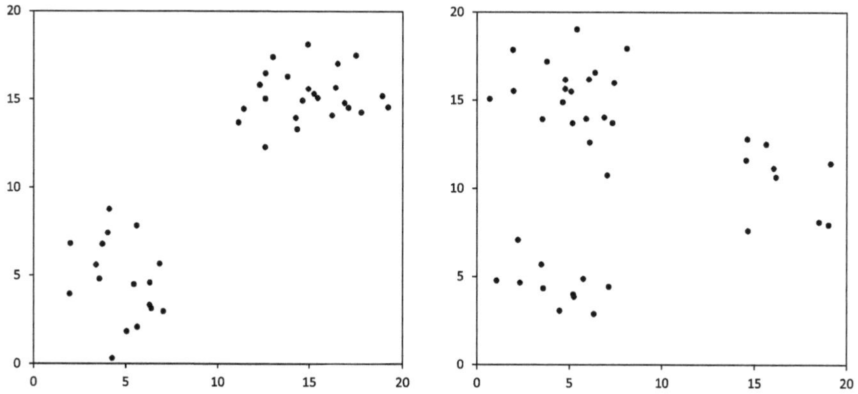

Fig. 11.1 Exemplary realization of $\mathscr{F}_{2C}$ and $\mathscr{F}_{3C}$ for $\mathscr{A}_{20}$

the outcome of the time budgeting. For these instances, ATB decision making might be impaired. We expect opposed results for AHS, because the anticipation drawing on spatial information given $\mathscr{F}_U$ may be vague. We further assume that the RAs may perform well for all distributions.

In the following, we define the spatial distribution functions for $\mathscr{A}_{20}$. All spatial parameters are depicted in kilometers. For $\mathscr{A}_{15}$, these parameters are reduced by factor 0.75. Given $\mathscr{F}_U$, a realization $l^C = (a_x, a_y)$ is defined as $a_x, a_y \sim U[0, 20]$. For $\mathscr{F}_{2C}$, the customers are equally distributed to each cluster. The cluster centers are located at $\mu_1 = (5, 5)$, $\mu_2 = (15, 15)$. The standard deviation within the clusters is $\sigma = 1$. The distribution is therefore point-symmetrical to the depot. For $\mathscr{F}_{3C}$, the cluster centers are located at $\mu_1 = (5, 5)$, $\mu_2 = (5, 15)$, $\mu_3 = (15, 10)$. 50% of the requests are assigned to cluster two, 25% to each of the other clusters. The standard deviations are set to $\sigma = 1$. Unlikely distribution outliers outside of $\mathscr{A}$ are tolerated and handled regarding Eq. 11.3.

11.2 Parameter Tuning

In this section, we describe the required tuning for the applied approaches. We start with the non-reactive approaches.

11.2.1 Non-reactive

The implicit anticipatory approaches WAS and WAE do not require any tuning. The explicit non-reactive anticipatory approaches CBH and AI are tuned as followed. To determine parameters d^*, κ for CBH, for every instance setting, we run 100 test runs within a learning phase. For every test run, we apply candidate parameter vectors (d_c^*, κ_c) with $d_c^* \in T$ and $\kappa_c \in [0, 2]$. We select the parameter combination maximizing the overall sum of confirmations for the execution phase. To determine the COG for AI, we sample a set of 100 spatial realizations $F \subset \mathscr{A}$ of the respective distribution $\mathscr{F}$.

11.2.2 Reactive

For all reactive anticipatory approaches, we neglect waiting to limit the number of PDSs. The detailed tuning of ATB, AHS, and the RAs is described in the following.

Anticipatory Time Budgeting

We tune AVI and the partitioning as proposed in Ulmer et al. (2017b). We consider $L = 5$ different levels of partitioning, starting with intervals of length 16 ($l = 5$) up to the discrete value representation ($l = 1$). The time and time budget consideration varies from 16 min down to minute by minute. We apply lookup tables with static interval length SLT(I) for all levels $I = 2^{l-1}, l = 1, \dots, 5$. For the weighted and dynamic LTs, we consider all five partitioning levels ($l = 5$ up to $l = 1$). We apply DLTs with $\zeta = 1.0, 1.25, \dots, 2.0$. For $\zeta = 1.0$, an entry is separated if its divergence in observations and standard deviation is above average. For $\zeta = 2.0$, the divergence has to be two times as high. For $\hat{V}_0$, we choose high initial values to force exploration in the beginning resulting in the selection of not yet observed entries. For all LTs, we run $|\bar{\Omega}| = 1$ million simulation runs. The derived policies are then used for the evaluation. To define the further specifics of the ADP algorithm, the interested reader is referred to the taxonomy provided by Powell (2007, p. 296ff). Powell differentiates seven characteristics to describe an ADP approach. For most of the characteristics, we maintain the "standard" design:

1. (c) State variable: The values are based on aggregated post-decision states.
2. (a, i) State sampling: States are sampled asynchronously. Pure exploitation is used.
3. (a) Average versus marginal values: The average state values are estimated.
4. (a) State space representation: The values are stored in a LT. We use static, weighted, and dynamic partitionings.
5. (b) Value selection for update: We update the values at the end of each simulation run.
6. (a) Number of samples: We update the values after each realization.
7. (a, i) Update of the values: We smooth the new estimate with the current approximation calculating the moving average.

Ad Hoc Sampling

In t, we run $n = 16$ sample runs with $m = \mathfrak{P}(n_t^T)$ sampled customers per run. The expected number of sampled customers therefore matches the expected number of customers between the current point of time t and time limit T. For sampling, we draw on the spatial distribution $\mathscr{F}$.

Rollout Algorithms

For both RAs X^m and X^a, we run $n = 16$ sample runs. The sampled customers follow the distribution Ξ. For the RA X^m with myopic base policy, no further tuning is necessary. For the hybrid RA X^a with ATB as base policy, we use DLT(N, σ) with $\zeta = 1.5$ and $|\bar{\Omega}| = 1$ million simulation runs.

11.3 Non-reactive Versus Offline Reactive

For the purpose of presentation, we first analyze the offline approaches to show the advantages of reactive compared to non-reactive anticipation. We analyze the impact of reactive offline anticipation on decision making in Sect. 11.4. We then compare online and offline approaches in Sect. 11.5. For ATB, CBH, the waiting approaches AI, WAS, WAE, and the myopic policy, we run 10,000 test runs for every instance setting and present the average percentage of served dynamic requests. The solution qualities for ATB drawing on DLT(N, σ) are shown in the upper part of Table 11.2. Since ATB is an offline approach, we can select the tuning providing the best solution quality (+). Additionally, we depict the average solution quality (∅) of the approach group. For each instance setting, the overall best result is printed in bold.

At first, we analyze the impact of the instance parameters on the solution quality. The number of served requests is dependent on the *dod*, the customer distribution, and the size of the service area. For varying *dod*, the number of served requests changes while the percentage remains constant. Given a small service area, more requests can be served due to shorter travel times. For the clustered distributions, customers are located close to each other and no far-off ERC reduce the initial time budget. In contrast to $\mathscr{F}_U$, insertion times are relatively low. In the extreme case of $\mathscr{A}_{20}$, $\mathscr{F}_U$, and expected 75 ERC (*dod* $= 0.25$), the initial tour already consumes the entire time budget. We exclude this instance in the following detailed analysis of the solution quality.

Comparing the solution approaches, the best results are achieved by ATB in 14 of the 17 remaining instance settings. ATB performs better given a large *dod* because the free time budget increases and the solution quality is less influenced by single customers and spatial information. CBH achieves a sufficient number of confirmations and anticipation regardless the instances, although it is generally not able to achieve the same solution quality as ATB. The explicit subset selection of CBH

Table 11.2 Served dynamic requests (in %)

Service area distribution		$\mathscr{A}_{15}$			$\mathscr{A}_{20}$		
		$\mathscr{F}_U$	$\mathscr{F}_{2C}$	$\mathscr{F}_{3C}$	$\mathscr{F}_U$	$\mathscr{F}_{2C}$	$\mathscr{F}_{3C}$
				dod = 0.25			
ATB	+	**50.1**	**87.4**	81.8	0.7	**66.5**	**55.2**
	∅	**50.1**	87.3	81.7	0.7	**66.5**	55.1
AI		44.4	**87.4**	82.7	0.7	65.5	51.4
WAE		45.0	**87.4**	**82.9**	0.7	65.9	51.7
WAS		14.1	40.0	34.0	0.3	24.5	16.9
CBH		47.7	87.2	82.6	**0.9**	66.2	53.3
myopic		44.6	80.5	79.5	0.7	64.3	50.8
				dod = 0.50			
ATB	+	**58.5**	**82.2**	74.1	**24.1**	**66.7**	**59.5**
	∅	58.4	81.4	73.9	**24.1**	66.5	59.4
AI		51.0	81.9	**75.2**	17.3	64.2	54.4
WAE		51.2	81.9	**75.2**	17.5	64.3	54.5
WAS		24.6	45.2	40.0	7.8	33.7	27.4
CBH		55.4	82.0	**75.2**	20.3	65.3	57.4
myopic		51.1	80.6	74.5	17.4	63.9	54.2
				dod = 0.75			
ATB	+	**59.6**	**78.6**	72.4	**45.4**	**66.2**	**59.8**
	∅	59.5	77.4	72.0	45.3	65.6	59.6
AI		53.5	77.7	72.5	35.4	62.5	56.2
WAE		53.7	77.7	72.5	35.6	62.5	56.3
WAS		34.1	50.7	46.6	21.4	40.1	36.3
CBH		57.1	77.7	**72.7**	41.2	63.5	58.2
myopic		53.6	77.4	72.3	35.5	62.4	56.2

avoids confirmations of inconvenient customers with high insertion times, but is not able to incorporate instance characteristics into decision making.

The results of waiting approaches depend on the determination of the points of time in which the vehicle idles. While WAS performs inferior to the myopic approach, AI and WAE allow anticipation and improve the results compared to the myopic approach up to 8.6%. The results for WAE always outperform AI. Hence, waiting at later points of time is more effective. Waiting approaches perform well if only a small number of dynamic customers is given. If the number of dynamic requests increases, waiting unnecessarily consumes time which cannot be used to serve (future) customer requests.

Table 11.3 Improvement (in %)

Service area distribution		$\mathscr{A}_{15}$			$\mathscr{A}_{20}$		
		$\mathscr{F}_U$	$\mathscr{F}_{2C}$	$\mathscr{F}_{3C}$	$\mathscr{F}_U$	$\mathscr{F}_{2C}$	$\mathscr{F}_{3C}$
				$dod = 0.25$			
ATB	+	12.3	8.6	2.9	0.0	3.4	8.7
	∅	12.3	8.4	2.8	0.0	3.4	8.5
AI		−0.4	8.6	4.0	0.0	1.9	1.2
WAE		0.9	8.6	4.3	0.0	2.5	1.8
WAS		−68.4	−50.3	−57.2	−57.1	−61.9	−66.7
CBH		7.0	8.3	3.9	28.6	3.0	4.9
				$dod = 0.50$			
ATB	+	14.5	2.0	−0.5	38.5	4.4	9.8
	∅	14.3	1.0	−0.8	38.5	4.1	9.6
AI		−0.2	1.6	0.9	−0.6	0.5	0.4
WAE		0.2	1.6	0.9	0.6	0.6	0.6
WAS		−51.9	−43.9	−46.3	−55.2	−47.3	−49.4
CBH		8.4	1.7	0.9	16.7	2.2	5.9
				$dod = 0.75$			
ATB	+	11.2	1.6	0.1	27.9	6.1	6.4
	∅	11.0	0.0	−0.4	27.6	5.1	6.0
AI		−0.2	0.4	0.3	−0.3	0.2	0.0
WAE		0.2	0.4	0.3	0.3	0.2	0.2
WAS		−36.4	−34.5	−35.5	−39.7	−35.7	−35.4
CBH		6.5	0.4	0.6	16.1	1.8	3.6

As assumed, in the cases of $\mathscr{A}_{15}$, $\mathscr{F}_{3C}$, AI and WAE are able to outperform ATB because of the varying sequence of clusters visited. This spatial information is not considered by ATB. Hence, the solution quality is impeded. Given the symmetrical $\mathscr{F}_{2C}$, the cluster sequence is not relevant for later outcomes allowing ATB to achieve a high solution quality without considering spatial information.

The percentage of improvement compared to the myopic approach is depicted in Table 11.3. ATB outperforms the myopic policy for nearly every instance. The best solutions gain an improvement of up to 38.5%. The average improvement over all 17 instance settings is 9.3%. Especially for $\mathscr{F}_U$, the average improvement of 20.9% is high because the explicit subset selection of ATB avoids customers far-off the current tour. Generally, in cases where the number of confirmations by the myopic approach is low, anticipation allows far better solutions, e.g., for $\mathscr{A}_{20}$ or $\mathscr{F}_U$. Here, expensive confirmations significantly influence the rest of the planning period. In cases, the number of accepted confirmations provided by the myopic solution is already high, the improvements are less significant. As a result, we can state that the reactive anticipation of ATB is advantageous for nearly every instance setting.

In the following, we analyze the reasons for the effective performance of the reactive approaches. We further show the advantages and disadvantages of offline and online approaches. First, we depict how reactive anticipation may allow an automatic capturing of the complex structure of problem and instances. Therefore, we analyze the offline approach ATB in detail in Sect. 11.4. Then, we compare the offline and online reactive approaches regarding runtime and solution quality in Sect. 11.5.

11.4 Offline Reactive Anticipation

In this section, we analyze the ATB algorithm. We compare the effectiveness and efficiency of the approximation process for ATB with DLT, WLT, and LTs with static equidistant interval lengths. For an exemplary instance, we analyze the structure of the DLT deducting statements about the achieved time budgeting decisions. We analyze the dependencies of tour duration, insertion times, expected values, and confirmation decisions over time.

We compare the results of the different partitioning approaches SLT, DLT, and WLT. The results are depicted in Table 11.4. The average results of DLT(N, σ) and DLT(N) are always higher compared to the SLTs. The average solution quality of the DLTs provides up to 15.9% more confirmations than the average SLTs and, in some cases, even outperforms the best SLT. Except for one case, DLT(N, σ) performs as well as or better than WLT. For 10 of the 17 instances, the best DLT(N, σ) outperforms WLT and the average DLT(N, σ) provides at least equal results. In the following, we analyze the different partitioning approaches regarding the approximation process and solution quality in detail. We especially focus on the DLTs.

Figure 11.2 shows the ATB-approximation process for SLT, WLT, and DLT(N, σ) given $\mathscr{F}_{2C}$, $dod = 0.75$, and $\mathscr{A}_{20}$ (Ulmer et al. 2017b). The x-axis represents the first 200,000 simulation runs. On the y-axis, the according solution quality is shown averaged over 10,000 simulation runs for a better presentation. We compare the tuned LTs providing the best solution quality after 1 million simulation runs and the average results of the tuning group. For this instance setting, DLT with $\zeta = 1.5$ and SLT with $l = 2$ achieve the best solution quality. Comparing DLT and SLT, the average DLT allows a faster approximation than the best SLT. For this instance, the best approximation in the first 10,000 simulation runs is achieved by WLT converging to similar results as the best DLT given 75,000 simulation runs.

By comparing the best and average solution quality of the SLTs over all instance settings, we can observe a gap of up to 15.8% in confirmations. As predicted, this confirms that the static and a priori definition of the interval size has a significant impact on the solution quality. Even though SLTs may allow approximation for one interval size, for others, the solution quality cannot exceed the myopic policy.

Table 11.4 Comparison of Partitionings, Served Dynamic Requests (in %)

Service area distribution		$\mathscr{A}_{15}$			$\mathscr{A}_{20}$		
		$\mathscr{F}_U$	$\mathscr{F}_{2C}$	$\mathscr{F}_{3C}$	$\mathscr{F}_U$	$\mathscr{F}_{2C}$	$\mathscr{F}_{3C}$
				$dod = 0.25$			
SLT	+	50.1	87.4	81.8	1.0	66.5	55.1
	∅	44.6	86.8	81.0	0.9	63.7	51.2
DLT(N, σ)	+	50.1	87.4	81.8	0.7	66.5	55.2
	∅	50.1	87.3	81.7	0.7	66.5	55.1
DLT(N)	+	50.1	87.3	81.8	0.8	66.5	55.1
	∅	50.0	87.3	81.7	0.8	66.4	55.1
DLT(σ)	+	49.7	87.4	81.7	1.0	66.6	54.9
	∅	47.4	87.2	81.1	0.9	65.5	52.2
WLT		50.0	87.3	81.7	1.0	66.5	55.1
				$dod = 0.50$			
SLT	+	58.4	81.9	74.3	24.1	66.2	59.5
	∅	54.2	79.1	71.0	20.8	64.2	57.6
DLT(N, σ)	+	58.5	82.2	74.1	24.1	66.7	59.5
	∅	58.4	81.4	73.9	24.1	66.5	59.4
DLT(N)	+	58.4	82.1	73.9	24.1	66.4	59.5
	∅	58.3	81.7	73.8	24.0	66.1	59.4
DLT(σ)	+	58.4	81.9	73.8	23.7	66.0	59.3
	∅	55.7	80.8	72.1	22.5	64.5	57.3
WLT		58.4	82.2	74.1	24.1	65.6	59.5
				$dod = 0.75$			
SLT	+	59.8	77.6	72.5	45.4	66.3	60.0
	∅	57.2	74.6	69.2	43.0	63.4	57.9
DLT(N, σ)	+	59.6	78.6	72.4	45.4	66.2	59.8
	∅	59.5	77.4	72.0	45.3	65.6	59.6
DLT(N)	+	59.8	77.9	72.2	45.4	65.7	60.1
	∅	59.7	77.5	72.0	45.2	65.1	59.8
DLT(σ)	+	59.8	77.5	72.2	45.2	66.0	59.6
	∅	57.6	76.6	69.6	43.7	63.3	58.4
WLT		59.0	77.9	72.5	45.4	65.8	59.4

The interval size providing the best solutions differs and depends on the number of simulation runs and instance settings. As expected, large interval lengths allow a fast approximation. Small interval lengths provide a high solution quality on the expense of a high number of required simulation runs. The best solutions are provided by $I = 1, 2, 4$. For $I > 4$, the detail of time-consideration is too low to provide a sufficient approximation. A comparison of the approximation behavior of SLT and DLT is depicted in Fig. 11.3 for $\mathscr{F}_{2C}$, $dod = 0.75$, and $\mathscr{A}_{20}$. Figure 11.3a shows the

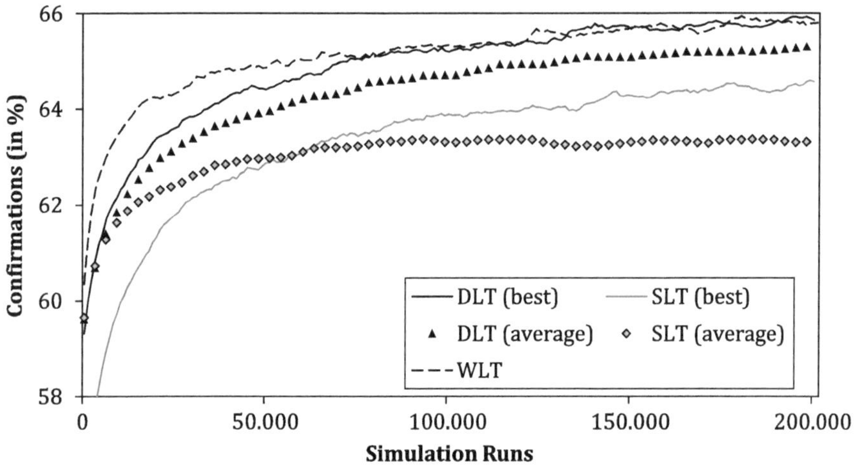

Fig. 11.2 Approximation process for $\mathscr{F}_{2C}$, $dod = 0.75$, and $\mathscr{A}_{20}$

development of the solution quality for SLTs with different interval sizes regarding the number of simulation runs. On the x-axis, the simulation runs are shown. On the y-axis, the according solution qualities of the average of 10,000 test runs are depicted. As we can see, SLT(2) and SLT(4) provide the best solution quality in the beginning. Then, SLT(4) provides a faster approximation process resulting in the best solution quality after 1 million simulation runs. After 5 million simulation runs, the best results are provided by SLT(2). Even though SLT(1) allows the highest detail, the results cannot outperform SLT(4) after 5 million simulation runs. Notably, the solution quality for SLT(16) decreases over the number of simulation runs. The partitioning is not detailed enough resulting in a defective evaluation of states. Compared to the SLTs, the DLTs show a homogenous behavior. The best results of the DLTs over all instances for 1 million simulation runs are achieved by $1.5 \leq \zeta \leq 2.0$. For $\zeta < 1.5$, the fast separation reduces the approximation speed slightly. The solution quality remains almost constant for the different ζ. The development for DLT(N) for different ζ is depicted in Fig. 11.3b. The solution quality only slightly differs regarding ζ. As expected, initially, relatively high ζ enable a fast approximation while for a higher number of simulation runs, lower ζ provide the best solution quality. Still, the difference between the different solution qualities is significantly lower compared to SLT. Table 11.4 shows that the DLT considering only one aspect differs in solution quality and approximation process. While DLT(N) achieves similar and for some instances even better results than DLT(N, σ), DLT(σ) is not able to provide high solution qualities for all instance settings. As expected, the separation of sparsely visited entries leads to an impaired approximation. Further, entries representing the beginning of the time horizon have a higher variance, because the stored values are higher compared to entries representing later times of the time horizon.

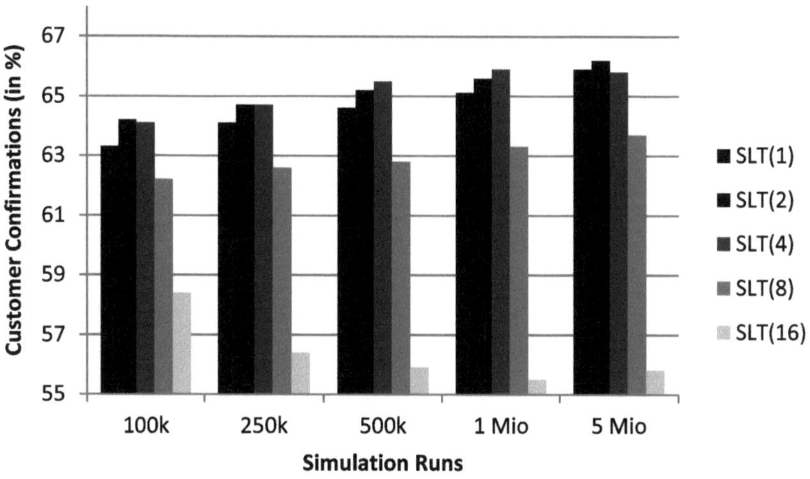

(a) SLT for Varying I

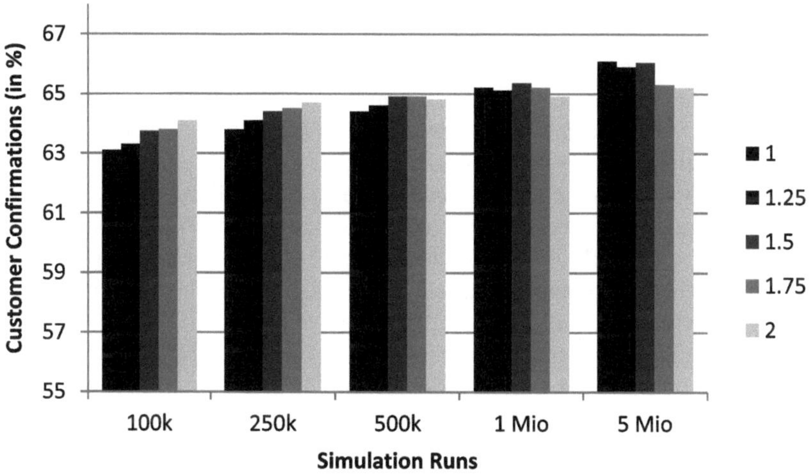

(b) DLT(N) for Varying ζ

Fig. 11.3 ATB solution quality over the number of simulation runs

In the following, we analyze the structure of the achieved DLTs. We exemplarily select the instance setting with $\mathscr{F}_{2C}$, $dod = 0.75$, and $\mathscr{A}_{20}$ because the instance structure allows a vivid display of the results and distinguished conclusions. Figure 11.4 shows the value development regarding the point of time for a fixed budget of $b = 100$. The values for are WLT dotted in bold. First, we analyze the development of the WLT-values. In the beginning with $t \leq 90$, the time budget generally exceeds 100 min. Hence, only a few observations are made. Between $90 \leq t \leq 120$, a plateau

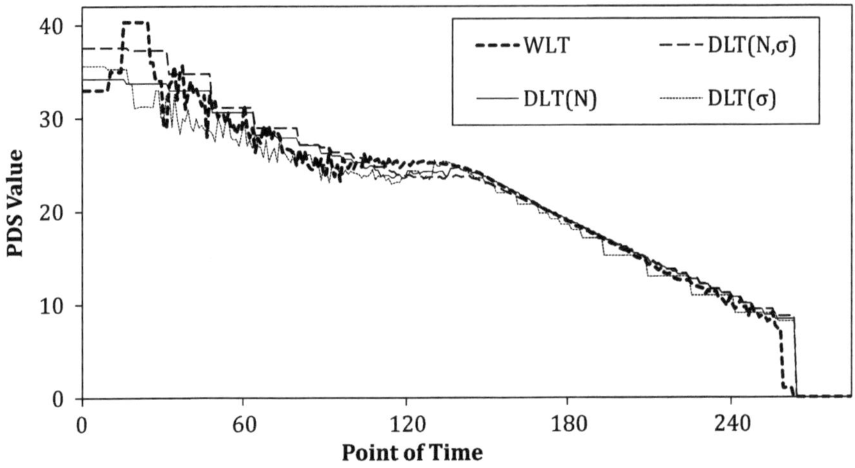

Fig. 11.4 Value over time given time budget $b = 100$ for $\mathscr{F}_{2C}$, $dod = 0.75$, and $\mathscr{A}_{20}$

in the value development can be observed and only a few decision points occur. The values remain constant between $90 \leq t \leq 120$ indicating a low number of confirmations within this time span. This results from the vehicle traveling between the two clusters as shown in the analysis of the routing in Sect. 11.4.1. For $120 \leq t \leq 260$, many observations and a continuous value decrease can be observed. Here, the vehicle serves customers in the second cluster. Due to the small travel times, the value decreases minute by minute. For $t > 260$, a budget of $b = 100$ is not feasible anymore. DLT(N, σ) and DLT(N) show an identical behavior compared to WLT with large intervals in the beginning and a detailed consideration in the following. DLT(σ) shows an opposed structure. This is explained in the following.

We now compare the structure of the DLTs in Fig. 11.5 after 10,000 simulation runs. Figure 11.5a shows the overall structure of DLT(σ). The x-axis represents the point of time, the y-axis represents the time budget. As already seen in Fig. 11.4, areas in the beginning of the time horizon are considered in detail and the entries in later areas remain unseparated. The relatively poor solution quality of DLT(σ) compared to the other DLTs indicates that a detailed consideration in the beginning may not be beneficial compared to a detailed focus in the course of the time horizon. The single large entry in the highly separated area results from the pure exploitation of the ADP algorithm. The first observation of this entry led to a low value prohibiting further observations. In contrast to DLT(σ), DLT(N) provides high quality solutions. The according LT-structure is shown in Fig. 11.5b. A correlation between point of time and free time budget over the time horizon can be identified. DLT(N) contains three segments, approximately divided at points of time $t = 120$ and $t = 240$. In these points of time, the separation level indicates a low number of observations. The time budget drops significantly right afterward. This reflects a heterogeneous observation

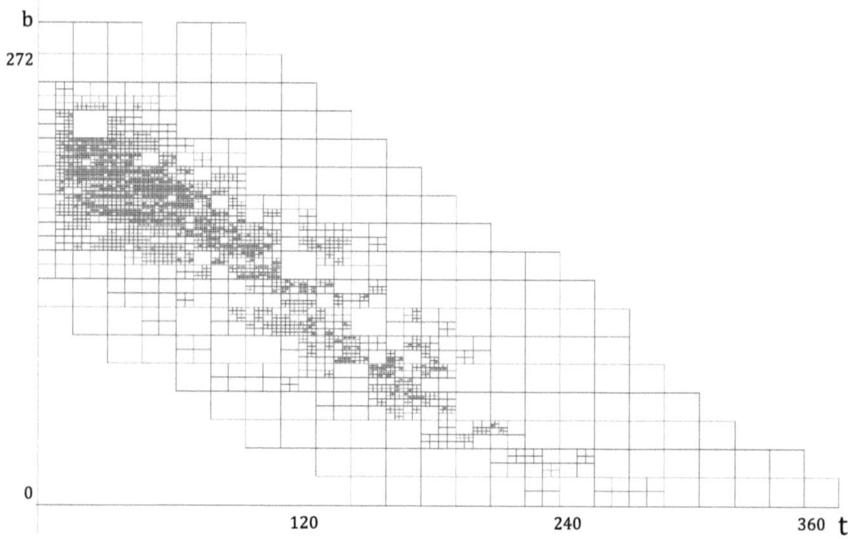

(a) DLT(σ) with $\zeta = 1.5$

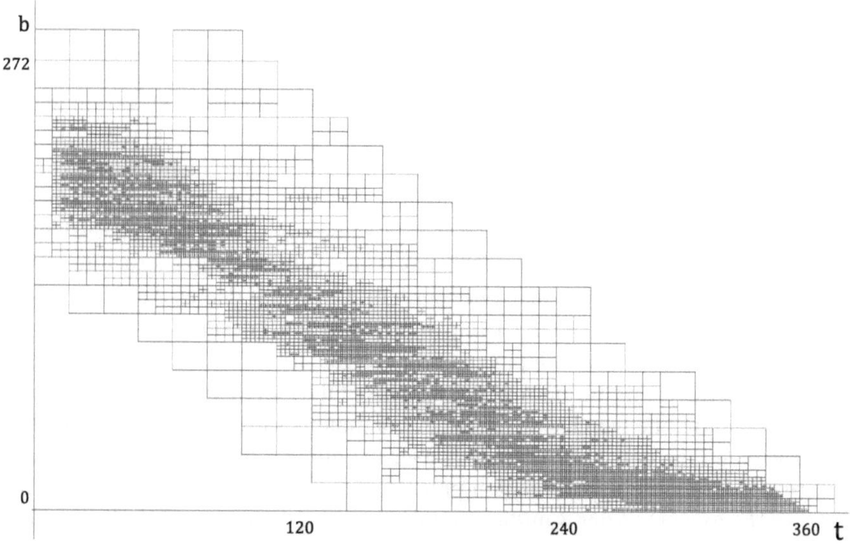

(b) DLT(N) with $\zeta = 1.5$

Fig. 11.5 Structure of DLTs for $\mathscr{F}_{2C}$, $dod = 0.75$, and $\mathscr{A}_{20}$ after 10,000 simulation runs

behavior over time. This behavior again results from routing. The segments are divided at times when the vehicle travels between clusters. By arriving in the new cluster, usually a large set of requests is given. The inclusions require a high amount of the free time budget leading to the drop in free time budget.

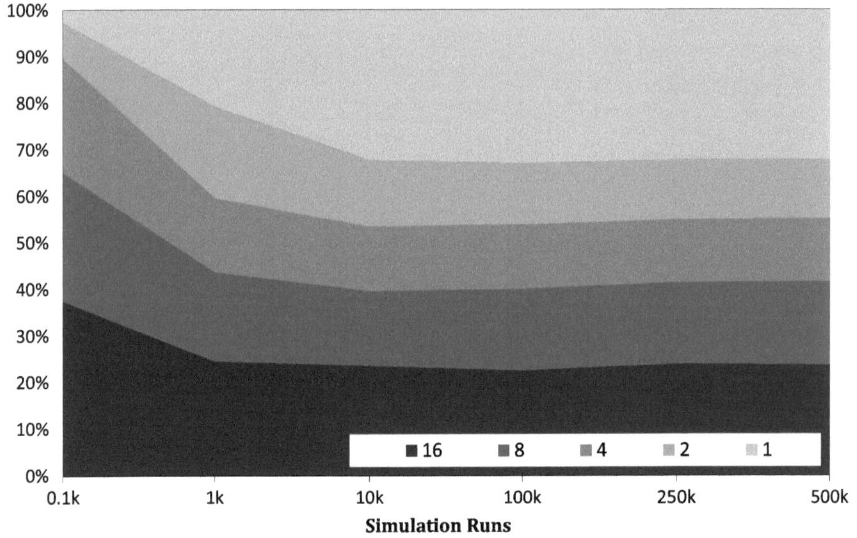

Fig. 11.6 Development of Entry-Sizes for DLT(N, σ) with $\zeta = 1.5$ given $\mathscr{F}_{2C}$, $dod = 0.75$, and $\mathscr{A}_{20}$

Figure 11.6 shows the percentage of entry-sizes over the first 500,000 simulation runs for DLT(N, σ) with $\zeta = 1.5$ for $\mathscr{F}_{2C}$, $dod = 0.75$, and $\mathscr{A}_{20}$. On the x-axis, the number of simulation runs is shown. On the y-axis, the according percentage for the different interval lengths is depicted. Initially, all entries have an interval-length of 16. During the approximation process, the percentage of entries with interval length of 1 constantly increases until the DLT has converged to a stable structure around 10,000 simulation runs. Less than one fourth of the table is depicted in highest level of detail. More than 30% of the LT remain in the original design.

Figure 11.7 shows the development of the weights for the different (sub-)tables of the WLT. On the y-axis, the normalized weights of the individual tables are depicted. In the beginning, the weights are equal. A continuous shift between the weights of the coarse and the detailed (sub-)tables over the number of simulation runs can be observed. During the approximation process, the main weight shifts subsequently from $w(16)$ to $w(1)$. After 3 million simulation runs, the main weight lies on $w(1)$ ending in $w(1) \geq 70\%$.

The number of entries of the different LTs after 1 million simulation runs is depicted in Table 11.5. For DLT, the average number of entries over parameter ζ is shown. For the given instance, WLT stores the values of $|\mathscr{E}(\text{WLT})| = 59{,}500$ entries after 1 million simulation runs, compared to $|\mathscr{E}(\text{DLT}(N, \sigma))| = 19{,}700$. The number of entries is dependent on the instance's characteristics. Given a large *dod* and, therefore, a high initial time budget, significantly more states can be reached resulting in partitionings with many separations. Again, the number of entries for

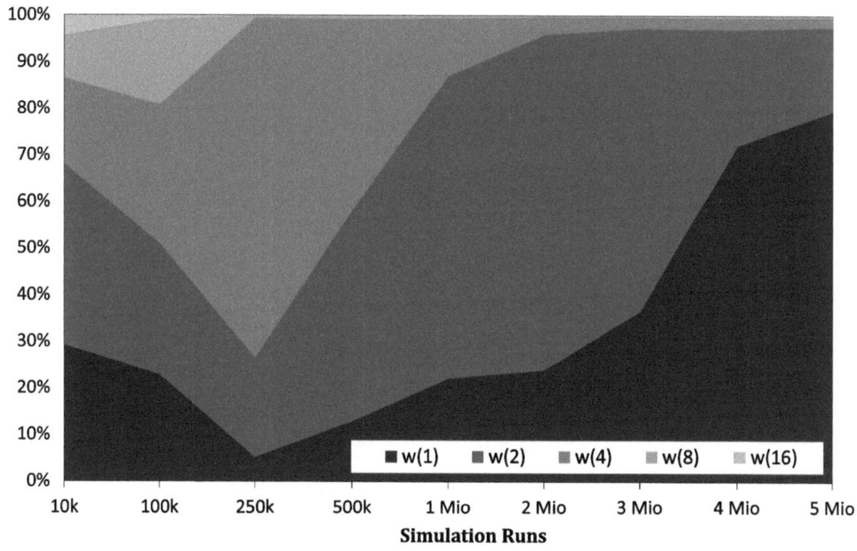

Fig. 11.7 Weights for WLT regarding the number of simulation runs

instance $\mathscr{F}_U$, $dod = 0.25$, and $\mathscr{A}_{20}$ is exceptionally small due to the high initial time budget consumption serving the ERC.

As expected, the number of entries of WLT is significantly higher than for the DLTs. The reduction of the number of entries compared to WLT is depicted in the lower part of Table 11.5. For the instance presented earlier, the required number of entries and the according memory consumption is reduced by 66.8%, Over all instances, this reduction reaches up to 68.1% for DLT(N, σ) and even up to 79.9% for DLT(N). Even though memory for storage of values may nowadays be broadly available, due to the exponential entry-increase by adding additional attributes to the LT, DLTs may be able to handle significantly higher-dimensional LTs compared to WLT.

11.4.1 Routing and Subset Selection

In the following, we analyze the decision policy achieved by ATB and the impact on subset selection and routing. We draw on the same instance presented earlier. To display the difference in confirmation behavior, Fig. 11.8a shows the average number of confirmations per point of time for ATB and the myopic approach over 10,000 test runs. Both approaches decide nearly similar in the beginning resulting in the same number of confirmations. This can be explained by observing the values regarding the free time budget in $t = 20$ as depicted in Fig. 11.8b. Between $200 \leq b \leq 260$, the values only slightly differ. This can be explained by the tradeoff between insertion

Table 11.5 Number of Entries per LT and Reduction of DLT Compared to WLT

Service area distribution	$\mathscr{A}_{15}$			$\mathscr{A}_{20}$		
	$\mathscr{F}_U$	$\mathscr{F}_{2C}$	$\mathscr{F}_{3C}$	$\mathscr{F}_U$	$\mathscr{F}_{2C}$	$\mathscr{F}_{3C}$
	Average number of entries (in 1,000)					
	$dod = 0.25$					
DLT(N, σ)	21.2	29.3	31.2	0.1	32.9	28.0
DLT(N)	15.0	22.2	23.2	1.2	25.5	20.6
DLT(σ)	21.7	23.8	25.0	8.0	23.8	25.0
WLT	47.7	60.2	64.8	20.7	60.5	54.2
	$dod = 0.50$					
DLT(N, σ)	24.1	22.6	25.3	12.8	29.5	29.9
DLT(N)	18.5	22.2	23.4	7.8	25.2	24.1
DLT(σ)	20.6	20.4	22.9	17.1	23.0	26.3
WLT	59.5	57.0	61.5	38.9	64.2	64.5
	$dod = 0.75$					
DLT(N, σ)	19.9	17.9	20.0	18.8	19.7	25.2
DLT(N)	19.0	19.6	20.9	13.7	19.7	22.6
DLT(σ)	19.9	17.5	18.3	21.1	19.8	23.0
WLT	62.5	52.7	58.6	59.1	59.5	65.2
	Reduction to WLT (in %)					
	$dod = 0.25$					
DLT(N, σ)	55.5	51.4	51.8	99.3	45.6	48.3
DLT(N)	68.6	63.1	64.2	94.3	57.9	61.9
DLT(σ)	54.5	60.5	61.4	61.4	60.6	53.9
	$dod = 0.50$					
DLT(N, σ)	59.5	60.4	58.8	67.1	54.2	53.7
DLT(N)	69.0	61.1	61.9	79.9	60.8	62.7
DLT(σ)	65.3	64.2	62.8	56.0	64.3	59.2
	$dod = 0.75$					
DLT(N, σ)	68.1	66.0	65.9	68.1	66.8	61.4
DLT(N)	69.6	62.9	64.3	76.9	66.9	65.3
DLT(σ)	68.2	66.7	68.8	64.3	66.7	64.7

times and area coverage as shown in Fig. 10.4. The increase of coverage counterbalances the insertion times leading to a nearly constant value. In the following hours of the shift, the myopic policy confirms more customers than ATB. As expected, the number of confirmations decreases drastically in the end of the shift.

Notable is the dent in confirmations for ATB in $t = 120$. This can be explained by the relatively low number of decision points, respectively, customer visits around this time. This results from the travel between the two customer clusters as assumed earlier and shown in Fig. 11.8c. Here, the number of customer visits per point of

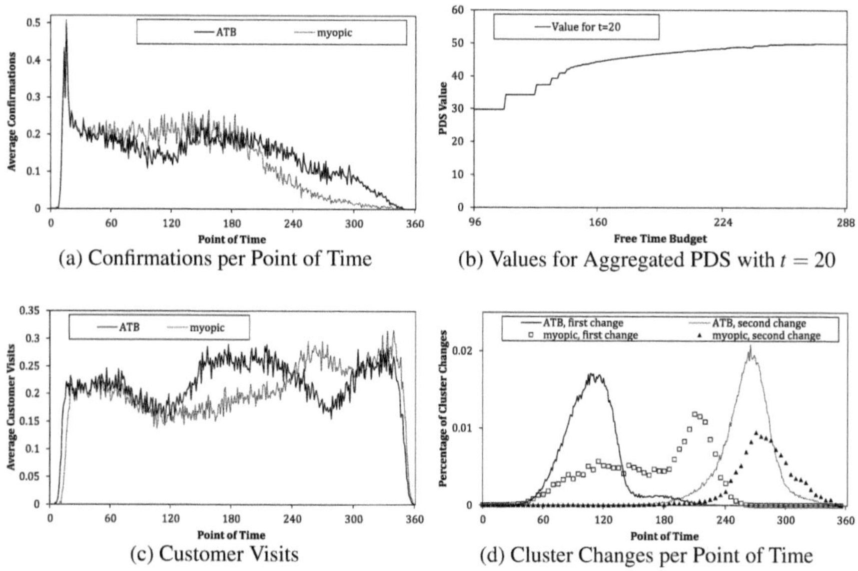

(a) Confirmations per Point of Time

(b) Values for Aggregated PDS with $t = 20$

(c) Customer Visits

(d) Cluster Changes per Point of Time

Fig. 11.8 Routing and confirmation properties for ATB and *myopic*

time is displayed. For ATB, three distinguished peaks of customer visits can be discovered. This can be explained by the vehicle changing clusters between the peaks. The points of time the vehicle changes clusters are depicted in Fig. 11.8d. The y-axis represents the percentage of test runs in which the vehicle leaves a cluster at a certain point of time. As expected, the times of the cluster changes for the myopic policy vary significantly. For ATB, the first cluster change mainly occurs between $100 \leq t \leq 125$. The average travel times between the two clusters of $\mathscr{F}_{2C}$ is $\tau = 20$. Generally, the vehicle arrives in the second cluster before $t = 146$ indicated by the increase in visits in Fig. 11.8c. The second cluster change occurs between between the $260 \leq t \leq 270$ and is even more distinct. The anticipatory confirmation policy significantly impacts the routing decisions and determines when to change clusters. This is remarkable because no explicit spatial information is included in the ATB state evaluation.

11.4.2 Budgeting Time

In this section, we analyze how ATB approaches the dependencies depicted in Fig. 10.3 and how ATB budgets the time accordingly. We compare ATB with CBH as approach with explicit subset selection and with *myopic* and WAS. We select WAS because the approach maximizes the coverage in the beginning. Figure 11.9 shows the impact of the approaches ATB, WAS, CBH, and *myopic* on the coverage and

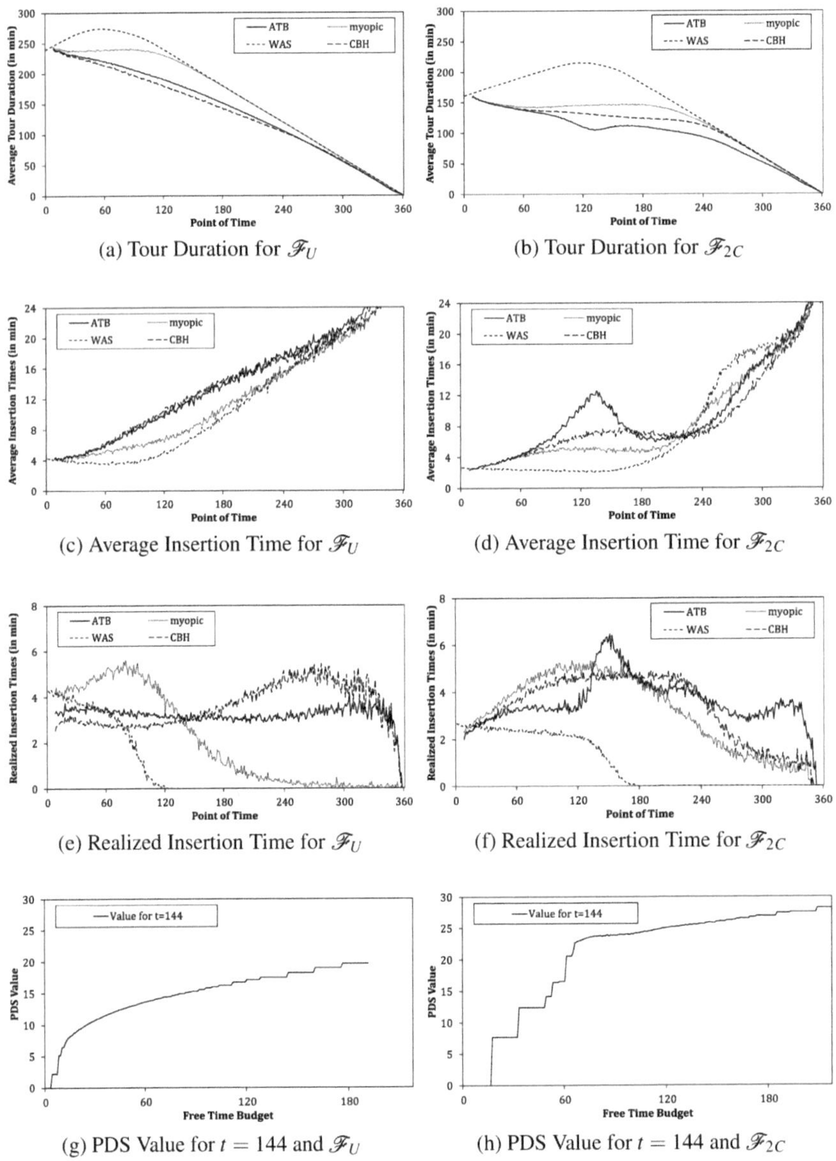

(a) Tour Duration for $\mathscr{F}_U$

(b) Tour Duration for $\mathscr{F}_{2C}$

(c) Average Insertion Time for $\mathscr{F}_U$

(d) Average Insertion Time for $\mathscr{F}_{2C}$

(e) Realized Insertion Time for $\mathscr{F}_U$

(f) Realized Insertion Time for $\mathscr{F}_{2C}$

(g) PDS Value for $t = 144$ and $\mathscr{F}_U$

(h) PDS Value for $t = 144$ and $\mathscr{F}_{2C}$

Fig. 11.9 Dependencies for the expected number of confirmations given $\mathscr{F}_U$ and $\mathscr{F}_{2C}$

insertion times and their decision making regarding the time budget for $dod = 0.75$ and $\mathscr{A}_{20}$. For ATB, we also show that the expected number of future confirmations depends on the free time budget for a specific point of time. The left side of Fig. 11.9 shows the results for $\mathscr{F}_U$, the right side for $\mathscr{F}_{2C}$ respectively. Figure 11.9a and b depict the development of the tour duration over time, Fig. 11.9c and d show the average insertion time γ_t per new request in decision point k with $t(k) = t$, calculated as depicted in Eq. 11.4. Decision x_a confirms all requests, regardless the time limit. γ_t^a is only calculated if new requests are given in decision point k: $|\mathscr{C}_k^r| > 0$.

$$\gamma_t^a = \frac{\bar{\tau}(\Theta_k^{x_a}) - \bar{\tau}(\Theta_k)}{\mathscr{C}_k^r} \tag{11.4}$$

Figure 11.9e and f depict the realized insertion time γ_t^c per confirmed customer and point of time as shown in Eq. 11.5. Decision x_c is the decision selected by the regarding approach. γ_t^c is only calculated if requests are confirmed: $|\mathscr{C}_k^c| > 0$.

$$\gamma_t^c = \frac{\bar{\tau}(\Theta_k^{x_c}) - \bar{\tau}(\Theta_k)}{\mathscr{C}_k^c} \tag{11.5}$$

Figure 11.9g and h show the development of the values regarding the free time budget for $t = 144$. We select this point of time because it lays within the arrival corridor of the vehicle in the second cluster for $\mathscr{F}_{2C}$. For some test runs, the vehicle arrives before $t = 144$. For other test runs, the vehicle arrives after $t = 144$.

Figure 11.9a and b show the average tour duration over time. As expected, approaches with explicit subset selection result in a shorter tour. WAS and myopic are able to even increase area coverage over time until the free time budget is expended. Notably, the ATB tour length increases after $t = 120$ for $\mathscr{F}_{2C}$. This is the time after the first change of clusters, when many new requests are confirmed. The tour length is strongly correlated to the actual area coverage indicated by the according insertion time in Fig. 11.9c and d. WAS and *myopic* have significantly lower average insertion times compared to ATB and CBH. The correlation of tour duration and insertion time is therefore strong. The average insertion time for $\mathscr{F}_U$ shows a continuous increase over time. For ATB and $\mathscr{F}_{2C}$, we can observe a peak around $t = 135$. The insertion time for requests around this point of time is high and falls for $t > 135$. At this point of time, the vehicle served all customers in the first cluster and arrived in the second. Hence, new requests within the first cluster are highly expensive to include. The same behavior can rudimentarily be seen for CBH. Considering the realized confirmations and the regarding insertion time in Fig. 11.9e and f, we can see that for $\mathscr{F}_U$, ATB on average allows the same amount of insertion time per customer over time. WAS and myopic allow a high amount in the beginning, but are not able to insert later requests because of the expended time budget. For $\mathscr{F}_{2C}$, a significant peak between $120 \leq t \leq 180$ can be seen. The highest realized insertion time is given around $t = 144$ meaning that at this point of time, the vehicle usually arrived in the second cluster. Around this time, ATB identifies the requirement to spend significantly more of the free time budget to reestablish the area coverage as already depicted in Fig. 11.9b.

As shown in Fig. 11.8d, the vehicle arrives in the second cluster usually before $t = 146$. We now analyze how this arrival impacts the values. Figure 11.9g and h show the value of a certain free time budget for $t = 144$ and DLT(N, σ). On the x-axis, the free time budget is shown. Given $\mathscr{F}_U$, the development of the values follows the idealized example of Fig. 10.7. An increase of the free time budget results in an increase of the value. Given $\mathscr{F}_{2C}$, a jump discontinuity can be seen at $b = 60$. For $b < 60$, the value drops drastically. A time budget lower than $b = 60$ reflects that the vehicle has already arrived in the second cluster. At arrival, a large subset of requests is confirmed. Hence, the post-decision state value drops. Given $b \geq 60$, the vehicle has not arrived in the cluster. A high number of confirmations can be expected and the value is significantly higher than for $b < 60$. ATB drawing on the DLT is able to identify this characteristic and, therefore, to adapt to the spatial and temporal request distribution.

11.5 Online Reactive Anticipation

In this section, we analyze the online reactive anticipatory approaches AHS and the RAs, X^m and X^a. First, we compare the online results with the offline ATB. Then, we analyze the online runtimes and how IZS may allow significant reductions of the RA-runtimes. Finally, we analyze the impact of hybrid anticipation and depict the advantages and disadvantages of spatial, respectively, temporal anticipation.

Table 11.6 Solution quality: confirmations (in %)

Service area	$\mathscr{A}_{15}$			$\mathscr{A}_{20}$		
dod	0.25	0.5	0.75	0.25	0.5	0.75
			$\mathscr{F}_U$			
ATB	51.3	57.8	60.0	0.4	22.5	46.5
AHS	50.4	56.9	60.1	0.5	21.3	44.4
X^m	50.5	56.2	58.8	0.5	21.2	43.7
X^a	51.2	58.2	60.8	0.5	22.5	46.4
myopic	47.2	50.6	54.0	0.4	16.2	36.9
			$\mathscr{F}_{2C}$			
ATB	86.9	82.4	76.1	66.7	64.9	66.7
AHS	86.8	83.9	77.9	67.6	66.7	67.7
X^m	87.1	83.6	76.9	68.4	67.0	67.2
X^a	87.0	84.4	76.9	68.2	67.5	68.8
myopic	87.0	83.1	76.5	66.2	62.8	63.6
			$\mathscr{F}_{3C}$			
ATB	81.0	74.4	71.7	53.5	59.4	60.0
AHS	81.9	76.8	74.0	53.3	59.5	60.8
X^m	82.8	77.5	73.2	53.9	59.8	61.5
X^a	82.9	77.8	73.7	54.3	61.4	62.6
myopic	81.8	75.4	72.1	49.2	55.2	56.2

11.5.1 Online Versus Offline Anticipation

In this section, we compare the advantages and disadvantages of online and offline approaches. As shown in Sect. 11.4, (offline) ATB is able to outperform the non-reactive anticipatory approaches for nearly every instance setting. Only in cases with small *dod* and for $\mathscr{F}_{3C}$, ATB is not able to provide high quality solutions. In cases with only a few requests and heterogeneous customer distribution, the spatial information of customer locations and routing may be important. This spatial information is considered in the online reactive approaches AHS and the different RAs X^m and X^a. We compare the results of the online approaches with ATB. Therefore, we run 250 test runs for every instance setting. The results are depicted in Table 11.6.

First, we compare AHS and ATB. For instances with a heterogeneous spatial distribution, AHS improves ATB-results by up to 3.1%. In these cases, the sampled customers are able to resemble the actual customers in the test run's realization. Especially for $\mathscr{F}_{3C}$, ATB is not able to explicitly anticipate cluster changes. As a result, ATB achieves in some cases even lower solution quality than the myopic policy. For uniformly distributed customers, an opposed behavior can be seen. AHS is not able to achieve similar results compared to ATB. For the instance with $dod = 0.5$ and $\mathscr{A}_{20}$, ATB improves the solution quality compared to AHS by 5.9%. For $\mathscr{F}_U$, the difference between sampled and realized customers is high. The spatial anticipation of ATB is ineffective. We analyze this behavior in more detail in Sect. 11.5.6.

We now analyze the behavior of the RAs. Compared to X^m, X^a achieves better results in 14 of the 18 instance settings leading to an improvement of up to 6.0%. The positive impact of the anticipatory base policy in X^a can therefore be verified. We analyze the benefits of combining offline and online anticipation of RA and ATB in Sect. 11.5.5. In comparison to the results of the according base policy, X^m improves the result on the myopic policy for every instance setting on average by 8.2% up to 30.8% for $\mathscr{F}_U$, $dod = 0.5$, and $\mathscr{A}_{20}$. X^a allows improvements to ATB in 16 of 18 cases. Neglecting the instance setting $\mathscr{F}_U$, $dod = 0.25$, and $\mathscr{A}_{20}$, the average improvement is 2.0%. This relatively low improvement can be explained by the already high quality decision making of ATB in many cases. The highest improvement is 4.6% given $\mathscr{F}_{3C}$, $dod = 0.5$, and $\mathscr{A}_{15}$. In this case, X^a enables the required explicit spatial anticipation that ATB is not capable of.

11.5.2 Runtime

AHS and the RAs are online approaches. In online decision making, the accessible calculation time in a decision point is limited. For every instance setting, we calculate the average runtime over all 250 realizations at the decision point of a realization with the highest required runtime. The approaches are implemented in Java 1.7.0_45 – b18. All calculations are conducted with an AMD Opteron Dual-Core Processor 2216 with 2.4 GHz and 8GB RAM, operating system Red Hat Enterprise Linux Server, Release 5.2. The according results are shown in Table 11.7. As

Table 11.7 Average maximal runtime (in seconds)

Service area	$\mathscr{A}_{15}$			$\mathscr{A}_{20}$		
dod	0.25	0.5	0.75	0.25	0.5	0.75
			$\mathscr{F}_U$			
ATB	0.0	0.0	0.0	0.0	0.0	0.0
AHS	1.8	16.4	129.3	0.0	9.6	125.7
X^m	0.2	1.9	39.8	0.0	2.0	85.3
X^a	0.6	3.7	46.7	0.0	5.3	113.5
myopic	0.0	0.0	0.0	0.0	0.0	0.0
			$\mathscr{F}_{2C}$			
ATB	0.0	0.0	0.0	0.0	0.0	0.1
AHS	1.8	13.8	88.7	2.0	17.6	175.6
X^m	0.2	2.4	50.9	0.3	5.2	410.5
X^a	0.4	3.7	69.6	0.6	6.0	247.1
myopic	0.0	0.0	0.0	0.0	0.0	0.1
			$\mathscr{F}_{3C}$			
ATB	0.0	0.0	0.0	0.0	0.0	0.0
AHS	1.9	12.3	83.2	1.9	19	174.2
X^m	0.2	1.4	14.5	0.2	2.7	84.8
X^a	0.5	2.3	29.1	0.6	4.4	96.4
myopic	0.0	0.0	0.0	0.0	0.0	0.0

expected, runtimes for ATB and the myopic policy are neglectable. The runtime for the online approaches increases with the *dod*. This is caused by the increasing number of sampled request. For AHS, each request has to be inserted in the existing tour via CI. For the RAs, each subset for a set of requests at every simulated decision point has to be evaluated.

The highest amount of runtime can be experienced for $\mathscr{F}_{2C}$ and $\mathscr{A}_{20}$. For these instance settings, a high number of requests is accumulated during the cluster change(s). The RAs have to simulate the cluster change for every subset of requests. Given $\mathscr{F}_{2C}$, $dod = 0.75$, and $\mathscr{A}_{20}$, X^m requires an average maximal calculation time of 410.5 s, i.e. 6.8 min. Assuming that operational decisions usually have to be made within a few minutes, the application of X^m may not be possible in this case. The indifference zone selection may reduce the required runtime. We analyze the application of IZS for this particular instance setting in Sect. 11.5.4. First, we show the development of ATB and the RAs by varying the number of sample runs.

11.5.3 Sample Runs

As expected, the solution quality and runtime increase with the number of sample runs. In this section, we analyze the behavior of the solution quality of the RAs and

Table 11.8 Confirmations and runtime regarding the number of samples

Distribution	$\mathcal{F}_U$			$\mathcal{F}_{2C}$		
n	AHS	X^m	X^a	AHS	X^m	X^a
			Confirmations (in %)			
2	43.9	39.5	44.3	67.3	60.8	64.1
4	43.9	40.9	45	67.6	63.3	65.2
8	44.1	42.1	45.8	67.8	65	67.2
16	44.4	43.7	46.4	67.7	67.2	68.8
32	44.6	44.7	46.7	67.6	68.4	69.4
64	44.4	45.1	46.9	67.6	69	70.1
128	44.5	45.3	47	67.6	69.4	70
			Runtime (in Minutes)			
2	0.3	0.2	0.2	0.4	0.6	0.6
4	0.5	0.7	0.4	0.8	0.9	1.1
8	1.1	0.8	0.7	1.5	1.8	2.5
16	2.1	1.4	1.9	2.9	6.8	4.1
32	4.4	2.5	3.7	6.0	5.4	8.1
64	8.3	5.0	7.6	11.7	25.3	14.9
128	16.4	11.3	11.2	23.3	36.3	31.4

AHS dependent on the number of sample runs. We exemplarily depict the results for instance settings $dod = 0.75$ and $\mathcal{A}_{20}$ for distributions $\mathcal{F}_U$ and $\mathcal{F}_{2C}$. We vary the number of sample runs between 2 and 128: $n = 2, 4, \ldots, 128$. Notably, the characteristics of a sample run for the RAs and AHS differ. For AHS, a sample run contains the sampling of a set of immediate requests at a specific decision point. For the RAs, a sample run results in the simulation of all subsequent decision points for a realization.

The confirmations are shown in the upper part of Table 11.8. The developments are depicted in Fig. 11.10. In Fig. 11.10a, the results for $\mathcal{F}_U$ are shown. In Fig. 11.10b, the results for $\mathcal{F}_{2C}$ are shown. On the x-axis, the number of sample runs is depicted. On the y-axis, the confirmations (in %) are shown. The diamonds depict the solution quality of the myopic policy, the triangles the solution quality of ATB. For $\mathcal{F}_U$, AHS only shows a slight improvement by an increase in sample runs. Here, the purely spatial anticipation is not able to achieve similar results compared to ATB even with a high number of sample runs. The RAs increase solution quality with an increase in sample runs and are eventually able to outperform their according base policy. Nevertheless, the results of X^a converge to the results of ATB. Given $\mathcal{F}_U$, temporal anticipation seems to be more suitable than spatial anticipation. For $\mathcal{F}_{2C}$, ATB performs relatively bad compared to the RAs and AHS. Again, an increase in sample runs does not improve the performance of AHS. Both RAs achieve the best solution quality, but X^a allows a faster increase and a higher solution quality in the end. This results from the offline anticipation induced by ATB. We analyze this impact in more detail in Sect. 11.5.5.

The according runtimes are depicted in the lower part of Table 11.8. For $\mathscr{F}_U$, the runtime increases nearly linear with the number of sample runs. This can be expected by the linear increase in required simulation runs. For AHS, the same development can be experienced for $\mathscr{F}_{2C}$. For the RAs, the runtimes show a different behavior. This can be explained by the changed decision policy indicated by the varying solution qualities. This decision policy influences the cluster change(s) and, therefore, significantly impacts the runtime. For AHS, solution qualities only slightly differ regarding the sample runs given $\mathscr{F}_{2C}$. Hence, the increase in runtime is linear.

As we can see, the runtimes for X^m and X^a increase up to more than 30 min given $n = 128$. Hence, the RAs with $n = 128$ are not applicable in operational decision making. In the following, we show how the same results can be achieved for X^a within

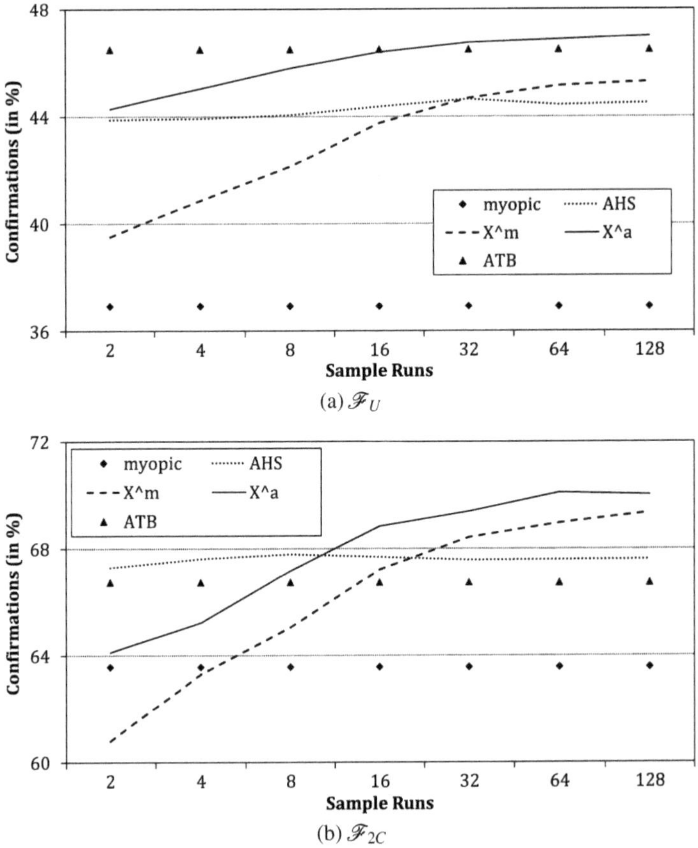

(a) $\mathscr{F}_U$

(b) $\mathscr{F}_{2C}$

Fig. 11.10 Development of confirmations for varying number of sample runs significant shorter runtime. Therefore, we apply the indifference zone selection as presented in Sect. 6.3.2.

11.5.4 Indifference Zone Selection

To reduce the number of sample runs, we use the fully sequential indifference zone selection procedure by Kim and Nelson (2001) to improve the performance of X^a. We focus on *dod* $= 0.75$, $\mathscr{A}_{20}$, and $\mathscr{F}_{2C}$ requiring the highest amount of runtime of all instance settings. We use the following parameter setting: We set the confidence interval to $\alpha = 0.01$, the indifference zone parameter to $\delta = 1.0$, and the maximal number of sample runs to $n_{\max} = 128$. $\delta = 1.0$ indicates that two post-decision states are indifferent if the average of both values is below 1.0. This is suitable, since the rewards in this problem are integer values. To analyze the behavior of IZS, we vary the number of minimal sample runs $n_0 = 2, 4, \ldots, 128$ for X^a. We run 250 test instances. The confirmations, runtimes, and average number of sample runs are shown in Table 11.9. We can experience an increase in solution quality, runtime, and average sample runs regarding an increasing n_0. This can be expected because an increase in n_0 results in a higher number of mandatory sample runs and therefore more reliable approximation. Further, IZS with $n_0 \geq 32$ usually terminates instantly after the initial n_0 runs. This termination results from the defined parameter setting of δ and α.

In the following, we compare the results with the regular X^a depicted in Table 11.8 to show the benefits of IZS. We are aiming for high solution quality and low runtimes. Hence, we have to consider two objectives: runtime and confirmations. To allow a comparison of IZS and regular X^a, we use tools of Pareto- or multiple criteria-optimization (compare Nemhauser et al. 1989, p. 665ff). We analyze the two approaches regarding the two objective dimensions. If an approach always requires more runtime for the same solution quality and provides a lower solution quality for the same runtime compared to a second approach, the first approach is called *dominated* by the second approach. In the following, we show that X^a based on

Table 11.9 Indifference zone selection

n_0	Confirmations (in %)	Runtime (min)	Sample runs
2	69.1	1.6	10.86
4	69.5	2.1	11.45
8	69.3	2.6	13.17
16	69.4	4.1	18.54
32	69.8	9.6	32.57
64	69.9	14.6	64.03
128	70.2	27.1	128.00

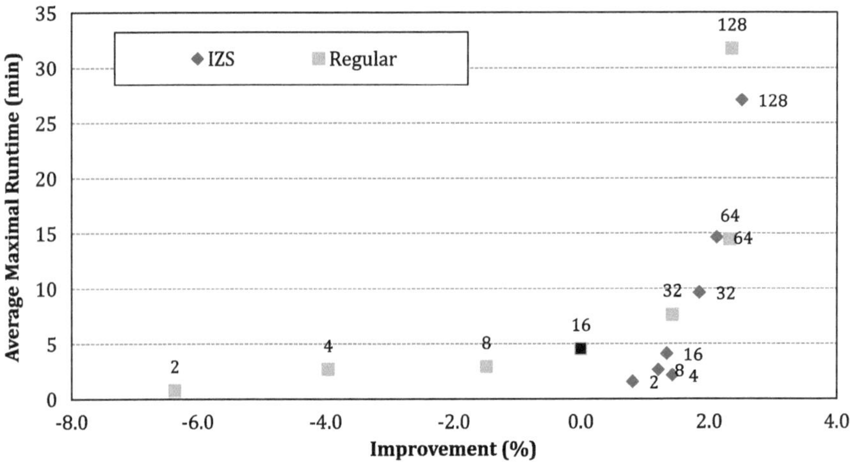

Fig. 11.11 Indifference zone selection for varying n_0, instance $\mathscr{F}_{2C}$, $dod = 0.75$, and $\mathscr{A}_{20}$

IZS dominates the regular X^a. Therefore, we depict the different instance settings in the two-dimensional objective space as shown in Fig. 11.11. We use regular X^a with $n = 16$ as benchmark. On the x-axis, the percentual improvement of the confirmations to the benchmark is shown. On the y-axis, the average maximal runtime is depicted. The diamonds depict the regular X^a, the squares X^a with IZS. Regular X^a with $n = 16$ is highlighted as benchmark.

Notably, IZS with $n_0 = 128$ and regular X^a with $n = 128$ result in similar solution approaches but in slightly different objective values and runtimes, even though they are applied to the same instance realizations. The difference can be explained by the stochastic samples within the algorithms. The same behavior can be seen for $n_0 = 64$ and partially for $n_0 = 32$. As shown in Table 11.9, IZS with $n_0 = 64$ only conducts 64.03 sample runs, IZS with $n_0 = 32$ only conducts 32.6 on average. For this problem and the selected IZS-parameters, IZS generally terminates after 32 runs regardless of the particular realization. For $2 \leq n_0 \leq 16$, we can experience a cluster of IZS with similar runtimes and solution qualities. The runtimes are all below the benchmark. The improvement is always positive. In this area of n_0, IZS shows a generic behavior allowing efficient and effective approximation. Even though the number of sample runs per decision point has to be equal or higher, the runtime of $n_0 = 16$ is lower than of the benchmark $n = 16$. This can be explained by a change in the decision making and therefore a different characteristic of the decision point with the maximal required runtime. In terms of Pareto-optimization, IZS spans a Pareto-front for $n_0 \leq 32$. This means that every individual parameter setting for regular X^a is dominated by a parameter setting of X^a with IZS. IZS allows an improvement of 1.43% compared to the benchmark reducing the runtime by 53.5%. In essence, IZS is able to efficiently improve the already high quality solutions of regular X^a.

To further analyze the behavior of IZS, we depict the average number of samples per point of time for $n_0 = 4$, $n_0 = 8$, and $n_0 = 16$ in Fig. 11.12. On the x-axis, the

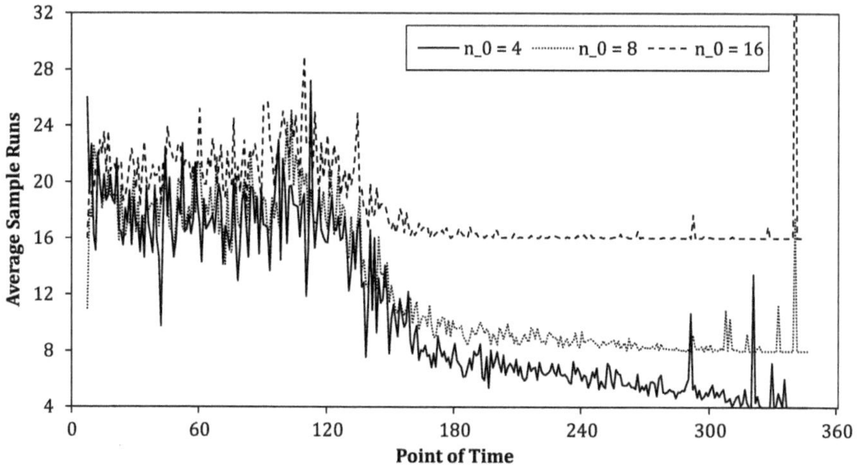

Fig. 11.12 Average number of sample runs over time for $n_0 = 4, 8, 16$

point of time is shown. On the y-axis, the average number of sample runs is depicted for decision points at the particular point of time. Because the number of test runs is 250 and therefore relatively low, a high noise and some outliers can be experienced. Still, a structure can be identified. In the beginning, the number of sample runs is relatively high and even increasing. At this point of time, the cluster change was not yet conducted. Hence, the RA has to simulate the cluster change. The realized values of post-decision states are significantly impacted by the cluster change and especially the set of sampled requests presented after the arrival in the second cluster. As shown in Sect. 11.4, the first cluster changes mainly happen in $100 \leq t \leq 150$. Hence, the number of required sample runs drops around $t = 120$. After $t = 180$, IZS with $n_0 = 16$ online conducts the mandatory 16 sample runs, while the number of sample runs for $n_0 = 4$ and $n_0 = 8$ still decreases. After the cluster change, a high number of sample runs is not required anymore.

11.5.5 Hybrid Anticipation

As mentioned earlier, the hybrid RA X^a allows to shift a percentage of the calculation effort into the offline approximation process of the base policy. This means that the locally detailed online anticipation can be partially substituted by the global low-dimensional anticipation. In the following, we show the impact of offline approximation on the resulting RA's policy. Therefore, we vary the number of simulation runs I of ATB and the number of sample runs n for X^a. We analyze $I =$ 0k, 1k, 10k, 100k, 1 million, and 5 million simulation runs. We differentiate $n = -, 2, 4, \ldots, 128$. $n = -$ indicates the immediate application of the base policy without the RA. As a result, the combination of $I = 0$ and $n = -$ results in the myopic policy. For every

Table 11.10 Offline versus online (Confirmations in %)

Runs (k)	–	2	4	8	16	32	64	128
0	63.6	61.3	62.9	65.3	67.1	68.4	68.9	69.3
1	57.3	58.9	60.7	63.1	65.3	67.2	68.8	69.3
10	61.6	60.4	62.5	64.8	66.9	68.7	68.9	69.7
100	65.9	63.2	64.8	66.7	68.0	68.8	69.6	69.7
1,000	66.7	64.3	65.9	67.3	68.8	69.7	70.0	70.0
5,000	66.8	64.7	66.0	67.6	68.8	69.6	69.7	70.3

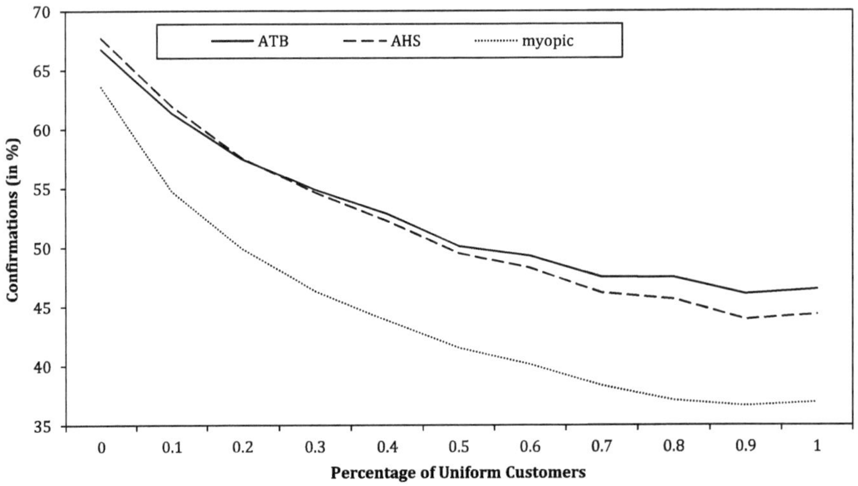

Fig. 11.13 Confirmations for varying parameter γ

combination, we run 250 sample runs. Table 11.10 shows the impact of offline simulation runs I of ATB and online simulation runs n of X^a for $\mathcal{F}_{2C}$, $\mathcal{A}_{20}$, $dod = 0.75$.

As expected, the solution quality increases with the number of simulation runs and sample runs. Only given $I = 1k$, $I = 10k$ or $n = 2$, $n = 4$, the approximated values of ATB are still weak resulting in a decrease in solution quality compared to the myopic policy. The combination of $I = 5{,}000k$ and $n = 128$ results in an improvement of 5.2% compared to plain ATB ($n = -$) and 1.4% compared to X^m ($I = 0k$). The shift of calculation effort to offline approximation allows a reduction of online sample runs of more than 75% comparing $I = 0$, $n = 128$ and $I = 1{,}000k$, $n = 16$, respectively, $I = 1{,}000k$, $n = 32$.

11.5.6 Spatial Versus Temporal Anticipation

In this section, we analyze the impact of the customer distribution to the spatial and temporal solution approaches, AHS and ATB. As seen in Sect. 11.5, spatial

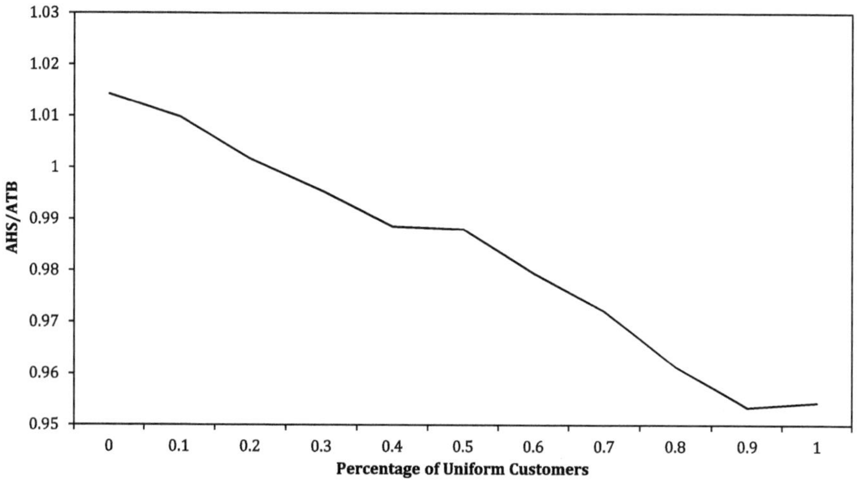

Fig. 11.14 AHS versus ATB for varying parameter γ

anticipation is advantageous if the customer locations are distributed in clusters. Temporal anticipation works better if the customers are uniformly distributed. To explain this behavior, we metaphorically draw on the idea of a Brownian motion with drift (Ross 2014, p. 612, BMD). BMD is a stochastic process containing a drift and a noise, i.e., Brownian motion. The drift describes the expected development of the BMD, the noise an undirected random factor. In context of the customer distributions, we can imagine a clustered distribution $\mathscr{F}_{2C}$ indicating a drift. The expected customer locations are relatively "reliable" compared to $\mathscr{F}_U$. The uniform distribution $\mathscr{F}_U$ can be seen as plain noise. To analyze the spatial and temporal anticipation in detail, we generate distributions $\mathscr{F}_\gamma$ with different percentages γ of drift and noise as depicted in Eq. 11.6.

$$\mathscr{F}_\gamma = (1-\gamma)\mathscr{F}_{2C} + \gamma\mathscr{F}_U \tag{11.6}$$

$\gamma = 0$ indicates no noise resulting in $\mathscr{F}_{2C}$. $\gamma = 1$ indicates no drift resulting in $\mathscr{F}_U$. We run 250 test runs for every distribution $\mathscr{F}_\gamma$ with $\gamma = 0.0, 0.1, \ldots, 1.0$. The resulting confirmations are depicted in Fig. 11.13. On the x-axis, γ is depicted. On the y-axis, the average confirmations for *myopic*, ATB, and AHS are shown. We can see that the confirmations significantly depend on γ. The explanation is twofold. First, the length of the initial tour and the insertion times increase with γ because more far-off customers have to be served. Second, the effectiveness of anticipation decreases with increasing noise and γ.

To compare the behavior of ATB and AHS in detail, we calculate the relation of the two solution qualities for changing γ. The resulting development is depicted in Fig. 11.14 (Ulmer et al. 2017a). On the x-axis, γ is depicted. On the y-axis, the confirmations of AHS divided by the confirmations of ATB are shown. We can see a continuous decrease given an increase in noise. Spatial anticipation is advantageous compared to temporal anticipation up to a noise of 20%. Generally speaking, the high presence of noise may require a more general, global anticipation while the presence of drift may allow and require detailed and local anticipation.

11.6 Implications

In this chapter and in the second part of the book, we have shown for an exemplary SDVRP with stochastic requests how reactive anticipation can be achieved and how it positively influences sequential decision making. To this end, we have applied, compared, and analyzed approaches of different degree of anticipation. We have highlighted the advantages of reactive anticipation compared to non-reactive anticipation. The reactive anticipatory approaches are able to achieve significantly better results adapting to specific instances. We have further shown the limitations and advantages of offline and online reactive anticipatory approaches and the potential of hybrid anticipatory methods. For the presented SDVRP with stochastic requests, hybrid RAs provide state-of-the-art decision making and are able to achieve outstanding anticipation in both the temporal and spatial developments during the process.

For the presented SDVRP, future work may focus on both the specific reactive anticipatory approaches and the problem setting. Until now, we are not able to achieve offline approximation explicitly considering spatial attributes in the AVI of ATB. Future work may identify suitable spatial and spatial-temporal attributes to include into ATB for an explicit consideration of customers' and vehicle's locations. The SDVRP may be extended to multi-vehicle and multi-periodical settings. Rejected customers might be postponed to a following period as presented by Angelelli et al. (2009). Here, anticipatory algorithms have to consider free time budgets for several vehicles and the current and the following periods.

Chapter 12
Conclusion and Outlook

In the final chapter of this book, we provide a brief summary, depict the main managerial implications of this book, and give an outlook on potentially fruitful areas of future research.

12.1 Summary

Practical routing applications contain a variety of different objectives, constraints, and drivers of uncertainty often resulting in the requirement for replanning or step-wise planning. The field of rich vehicle routing aims on modeling and approaching these real-world problems to provide quantitative decision support. In this book, we have focused on a specific aspect of RVRPs, the anticipation of uncertainty and future replanning in current decision making. The according problems can be described as stochastic dynamic vehicle routing problems. In the first part of the book, we have identified travel times, service times, customer requests, and customer demands as the four main drivers of uncertainty. We have shown how SDVRPs can be modeled as MDPs and highlighted the MDP's suitability to consider subsequent decision making and uncertainty. We have further depicted how stochastic information derived by predictive analytics can be integrated in the solution approach for the according SDVRP. We have presented the concept of anticipation and classified anticipation regarding the consideration of the problem's potential reactions and the integration of stochasticity. Based on the classification, we gave an extensive literature review regarding the work on SDVRPs.

For reactive anticipation, we have described a set of general methods of approximate dynamic programming and extended two ADP-methods to allow efficient and effective approximation. For offline anticipation, the dynamic lookup table adapts the state space partitioning according to the approximation process. The DLT enables a

© Springer International Publishing AG 2017

M.W. Ulmer, *Approximate Dynamic Programming for Dynamic Vehicle Routing*,
Operations Research/Computer Science Interfaces Series 61,
DOI 10.1007/978-3-319-55511-9_12

problem-specific approximation process and significantly reduces the required simulation runs as well as the memory for the storage of values. First results for a multi-periodical SDVRP with stochastic customer requests show that for a four-dimensional vector space the DLT is additionally able to significantly improve the overall solution quality compared to equidistant LTs (Ulmer et al. 2016b). For online anticipation, we proposed to combine rollout algorithms with the indifference zone selection. The combination improves both the achieved policy and the computational performance of the algorithm. Finally, we have described how to combine offline AVI and online RA to achieve hybrid anticipation exploiting the advantages of offline and online anticipation.

In the first part of the book, we have identified stochastic customer requests as the most prominent driver of uncertainty in both real-world applications and scientific work. In the second part of the book, we have presented an exemplary routing application with uncertainty in customer requests to show the scientific process originating from the practical application ending in the analysis of reactive anticipation's impact. We have modeled the real-world problem as SDVRP via a Markov decision process and presented a variety of solution approaches based on the literature. We have further shown the process to achieve approaches providing reactive anticipation in detail. We have transferred the general ADP-methods from the first part of the book to the specific SDVRP. In extensive computational studies, we have highlighted the advantages of reactive anticipation. We could show that the combination of AVI based on DLT and RA drawing on IZS leads to a significant improvement of the approximation process and the achieved policy. Further, we could confirm that RAs can be seen as a tool for improvement of (nearly) every base policy (compare Ulmer et al. 2017a). We have been finally able to show that a hybrid RA is able to combine the advantages of offline and online reactive anticipatory approaches. The hybrid RA achieves by far the highest solution quality and outperforms state-of-the-art benchmark heuristics from the literature significantly.

12.2 Managerial Implications

In the following, we depict the book's main managerial implications for logistic service providers.

- Due to developments like digitization, urbanization, and increasing customer expectations, logistic service providers experience a substantial percentage of the information relevant for planning only revealing subsequently over time, i.e., uncertainty.
- The impact of handling uncertainty on the success of service providers' decision making is already significant and will further increase in the future. A consideration of uncertainty in decision making becomes therefore inevitable.
- Logistic service providers have access to vast amounts of historical data. This allows predictive analytics, i.e., the derivation of stochastic information for routing

applications. The increasing available computational resources enable immediate adaptions of plans due to newly revealed information, i.e., dynamic decision making.
- Efficient and effective planning requires the integration of uncertainty, i.e., the stochastic information and potential future adaptions of plans, i.e., subsequent decision making in the applied decision support methods. This is achieved by anticipatory planning methods or prescriptive analytics respectively.
- The explicit integration of stochastic information and subsequent decision making is achieved by reactive anticipation. Approaches drawing on reactive anticipation are able to adapt to the structure of the specific problem as shown by means of an exemplary SDVRP in the second part of this book. As a result, these approaches are significantly more suitable compared to conventional approaches only providing non-reactive anticipation.
- Reactive anticipation is generally achieved by methods of approximate dynamic programming evaluating the outcome of current decisions regarding the expected future rewards or costs respectively.
- We can distinguish online and offline ADP-methods both having advantages and shortcomings. Online approaches allow limited, but detailed consideration of the plan's impact to the future developments. Offline approaches enable a global consideration of the problem, but only on an aggregated level of detail. The hybrid combination of online and offline ADP-methods is possible and enables an alleviation of the shortcomings and the exploitation of the advantages.

With these deductions at hand, we can give directions for future research in the field of SDVRPs and anticipation.

12.3 Future Research

In the future, we see requirements for both research in anticipation for specific application fields of SDVRPs and research regarding the general methodology to achieve reactive anticipation for SDVRPs.

12.3.1 Application Fields

Our literature review indicates that reactive anticipation and dynamic decision making are achieved for stochastic *demands* and *requests* but are not yet fully established given uncertainty in *service times* and *travel times*. In reference to urbanization and demographic developments, these field may become more important and the percentage of uncertainty in these fields may increase in the future. Anticipatory approaches may allow facing the emerging challenges. Reactive anticipation may be able to generate new or improve existing routing approaches. As an example, stochastic travel times in the city often follow a spatial-temporal distribution. Here, the spatial and temporal anticipation provided by the RAs may be particularly suitable. In Chap. 3, we identified that the percentage of uncertainty is high especially in the fields of emergency routing, demand responsive passenger transport, shared mobility, courier services, parcel pickups, and same-day delivery. Here, anticipation may be in particular necessary for efficient and effective planning and these applications may significantly benefit from the development of anticipatory solution approaches. We further identified real-world routing applications in the fields of technician, healthcare, and salespersons routing, patient transport, delivery of groceries and retail worthwhile to be approached by anticipatory solution approaches. Future research may further focus on prescriptive analytics in these fields of application.

12.3.2 Reactive Anticipation for SDVRPs

As shown in this book, reactive anticipation is generally achieved by methods of approximate dynamic programming. To allow reactive anticipation for SDVRPs, these general online and offline ADP-methods have to be tailored to the specific application. Nevertheless, we can state some general points ADP-methods require for the application to SDVRPs. These points consider the state space representation and the method itself for offline and online anticipation described in the following. We finally propose how reactive anticipation may be tailored to the practical requirements of LSPs.

In offline anticipation, the state space, decision space, and transition space have to be represented efficiently for a tractable approximation process and effectively for the approximation's quality. For many SDVRPs, the state, decision, and transition spaces are vast and the *curses of dimensionality* significant. As depicted in this book, especially the state spaces are generally complex for SDVRPs and the state space representation essential for a successful offline anticipation. States contain many temporal and spatial dimensions. A detailed consideration of every aspect may not be computationally tractable. Hence, future research has to identify suitable state space representations for the specific SDVRPs. Further, the offline ADP-methods have to enable *adaptions* of the state space regarding the structure of the SDVRPs revealed during the approximation process. Methods have to enable adaptive reductions and aggregations of the state space regarding the approximation process. State attributes may be added or removed as proposed by George et al. (2008) dependent on the quality of value approximation for a specific state. Generally, an iterative exchange between prescriptive and predictive analytics as proposed in Meisel and Mattfeld (2010) may be promising to reveal the important problem features and incorporate them in the approach. Further, instead of aggregation techniques, an adaptive clustering of states to representatives may avoid the loss of detail but still enable efficient approximation (Lee and Lau 2004; Hamagami and Hirata 2005). In these cases, states may function as representatives of clustered sets of states and the clustering may be adapted regarding the approximation process (Soeffker et al. 2016).

For online methods, approximation has to be efficient because the runtime is limited. Future work may aim on further reducing the required runtime of online methods. Until now, online methods still lack the possibility to explicitly incorporate the global features of the problem. Here, an increasing level of detail within the online approximation may allow to add general global features to the detailed local consideration. The concept of hybrid anticipation may be further developed in the field of SDVRPs. As an example, the combination of offline and online methods may be iteratively applied. The online method may feedback the achieved approximation to the offline method. This may enable an adaptive hybrid approach and a seamless merging of offline and online anticipation.

Finally, ADP-methods may additionally be aligned to realistic practical requirements of the logistic service providers. As an example, LSPs may aim on reliable and less volatile outcomes of the future, i.e., LSPs may be *risk-averse* (Dyer and Sarin 1982; Jiang and Powell 2015; Ulmer and Voß 2016). Generally, ADP-methods aim on estimating the (risk-neutral) expected future mean outcomes. Heterogeneous volatilities of realized values for different states are not considered. Risk-aversion in dynamic decision problems is challenging because the rewards accumulate over subsequent decision points (Ruszczyński 2010). A straightforward consideration of states' value variances is not applicable. In other cases, predictions and stochastic information may be distorted (George and Powell 2007) or may not be able to capture the exact properties of the real-world uncertainty. As an example, uncertain travel times are, amongst others, influenced by street loads, congestion, and traffic management on several street segments (Ehmke and Campbell 2014). In combination with predictive analytics, it may be worthwhile analyzing how *inaccurate information* and the *reliability of information* can be anticipated.

Concluding this book, we can state that uncertainty and complexity of the routing problems as well as the amount of accessible data and computational resources will further grow demanding suitable solution approaches from the field of prescriptive analytics. Conventional myopic or non-reactive anticipatory approaches are limited and will not be able to master these developments sufficiently. We are confident that reactive anticipation is the key for future success in stochastic dynamic vehicle routing.

References

Adulyasak, Yossiri, and Patrick Jaillet. 2015. Models and algorithms for stochastic and robust vehicle routing with deadlines. *Transportation Science* (online available).

American Public Transportation Association. 2013. Number of unlinked passenger trips in demand responsive transport in the united states from 1996 to 2013 (in 1,000s). http://www.statista.com/statistics/205293/unlinked-passenger-trips-in-us-demand-responsive-transport-since-1996/. Accessed 21 Sept 2015.

Angelelli, Enrico, Nicola Bianchessi, Renata Mansini, and Maria G. Speranza. 2009. Short term strategies for a dynamic multi-period routing problem. *Transportation Research Part C: Emerging Technologies* 17 (2): 106–119.

Barceló, Jaime, Hanna Grzybowska, and Sara Pardo. 2007. Vehicle routing and scheduling models, simulation and city logistics. *Dynamic Fleet Management*, 163–195. Berlin: Springer.

Barr, Alistair, and Jessica Wohl. 2013. Exclusive: Wal-mart may get customers to deliver packages to online buyers. *Reuters*. http://www.reuters.com/article/2013/03/28/us-retail-walmart-delivery-idUSBRE92R03820130328. Accessed 28 Aug 2015.

Barto, Andrew G. 1998. *Reinforcement Learning: An Introduction*. Cambridge: MIT Press.

Beaudry, Alexandre, Gilbert Laporte, Teresa Melo, and Stefan Nickel. 2010. Dynamic transportation of patients in hospitals. *OR Spectrum* 32 (1): 77–107.

Behrmann, Elisabeth, and Richard Weiss. 2014. Volvo said to near deal to deliver parcels to parked cars. http://www.bloomberg.com/news/articles/2014-11-17/volvo-said-to-near-deal-to-deliver-parcels-to-parked-cars. Accessed 06 May 2015.

Bektas, Tolga, Panagiotis P. Repoussis, and Christos D. Tarantilis. 2014. Dynamic vehicle routing problems. *Vehicle Routing: Problems, Methods, and Applications*, 299–348. Philadelphia: Society for Industrial and Applied Mathematics.

Bellman, Richard. 1956. Dynamic programming and Lagrange multipliers. *Proceedings of the National Academy of Sciences of the United States of America* 42 (10): 767.

Bent, Russell, and Pascal Van Hentenryck. 2007. Waiting and relocation strategies in online stochastic vehicle routing. *IJCAI* 1816–1821.

Bent, Russell W., and Pascal Van Hentenryck. 2003. Dynamic vehicle routing with stochastic requests. *IJCAI* 1362–1363.

Bent, Russell W., and Pascal Van Hentenryck. 2004. Scenario-based planning for partially dynamic vehicle routing with stochastic customers. *Operations Research* 52 (6): 977–987.

Beraldi, Patrizia, and Maria E. Bruni. 2009. A probabilistic model applied to emergency service vehicle location. *European Journal of Operational Research* 196 (1): 323–331.

Berbeglia, Gerardo, Jean-François Cordeau, and Gilbert Laporte. 2010. Dynamic pickup and delivery problems. *European Journal of Operational Research* 202 (1): 8–15.

© Springer International Publishing AG 2017

M.W. Ulmer, *Approximate Dynamic Programming for Dynamic Vehicle Routing*, Operations Research/Computer Science Interfaces Series 61,
DOI 10.1007/978-3-319-55511-9

Berthold, Michael, and David J. Hand. 2003. *Intelligent data analysis: an introduction*. Berlin: Springer Science & Business Media.

Bertsekas, Dimitri P., and David A. Castañon. 1989. Adaptive aggregation methods for infinite horizon dynamic programming. *IEEE Transactions on Automatic Control* 34 (6): 589–598.

Bertsekas, Dimitri P., and John N. Tsitsiklis. 1996. *Neuro-dynamic programming*. Belmont: Athena Scientific.

Bertsekas, Dimitri P., John N. Tsitsiklis, and Cynara Wu. 1997. Rollout algorithms for combinatorial optimization. *Journal of Heuristics* 3 (3): 245–262.

Bertsimas, Dimitris, and Ramazan Demir. 2002. An approximate dynamic programming approach to multidimensional knapsack problems. *Management Science* 48 (4): 550–565.

Bertsimas, Dimitris J., and Garrett Van Ryzin. 1991. A stochastic and dynamic vehicle routing problem in the Euclidean plane. *Operations Research* 39 (4): 601–615.

Bianchi, Leonora. 2000. Notes on dynamic vehicle routing-the state of the art. Technical report, Istituto Dalle Molle Di Studi Sull Intelligenza Artificiale.

Black, Fischer, and Myron Scholes. 1973. The pricing of options and corporate liabilities. *The Journal of Political Economy* 637–654.

Blum, Christian, and Andrea Roli. 2003. Metaheuristics in combinatorial optimization: Overview and conceptual comparison. *ACM Computing Surveys (CSUR)* 35 (3): 268–308.

Branke, Jürgen, Martin Middendorf, Guntram Noeth, and Maged Dessouky. 2005. Waiting strategies for dynamic vehicle routing. *Transportation Science* 39 (3): 298–312.

Brinkmann, Jan, Marlin W. Ulmer, and Dirk C. Mattfeld. 2015. Short-term strategies for stochastic inventory routing in bike sharing systems. *Transportation Research Procedia* 10: 364–373.

Bubner, Nedialka, Nikolaus Bubner, Ralf Helbig, and Martin Jeske. 2014. Logistics trend radar. Technical report, DHL. http://www.dhl.com/content/dam/downloads/g0/about_us/logistics_insights/DHL_Logistics-TrendRadar_2014.pdf. Accessed 27 Aug 2015.

Burger, Marie-Susann, Birte Janßen, Felix Kapel, Wenke Krützfeldt, Jens Langner, and Volker Staffa. 2015. Industry report - truck transportation. Technical report. Statista.

Butz, Martin V., Olivier Sigaud, and Pierre Gérard. 2003. Anticipatory behavior: Exploiting knowledge about the future to improve current behavior. *Anticipatory Behavior in Adaptive Learning Systems*, 1–10. Berlin: Springer.

Caceres-Cruz, Jose, Pol Arias, Daniel Guimarans, Daniel Riera, and Angel A. Juan. 2014. Rich vehicle routing problem: Survey. *ACM Computing Surveys* 47 (2): 32:1–32:28.

Campbell, Ann M. 2006. Aggregation for the probabilistic traveling salesman problem. *Computers and Operations Research* 33 (9): 2703–2724.

Campbell, Ann M., and Martin Savelsbergh. 2006. Incentive schemes for attended home delivery services. *Transportation Science* 40 (3): 327–341.

Campbell, Ann M., Michel Gendreau, and Barrett W. Thomas. 2011. The orienteering problem with stochastic travel and service times. *Annals of Operations Research* 186 (1): 61–81.

Capgemini. 2012. Number of global e-commerce transactions from 2011 to 2015 (in billions). http://www.statista.com/statistics/369333/number-ecommerce-transactions-worldwide/. Accessed 21 Sept 2015.

Cartledge, James. 2012. Belgian firm to deliver ecommerce packages to parked cars. http://postandparcel.info/51790/news/it/belgian-firm-to-deliver-ecommerce-packages-to-parked-cars. Accessed 06 May 2015.

Chardy, Matthieu, and Olivier Klopfenstein. 2012. Handling uncertainties in vehicle routing problems through data preprocessing. *Transportation Research Part E: Logistics and Transportation Review* 48 (3): 667–683.

Chen, Xi., Barrett W. Thomas, and Mike Hewitt. 2015. Multi-period technician scheduling with experience-based service times and stochastic customers. Technical report, University of Iowa.

Chen, Zhi-Long, and Hang Xu. 2006. Dynamic column generation for dynamic vehicle routing with time windows. *Transportation Science* 40 (1): 74–88.

Christiansen, Christian H., and Jens Lysgaard. 2007. A branch-and-price algorithm for the capacitated vehicle routing problem with stochastic demands. *Operations Research Letters* 35 (6): 773–781.

Christofides, Nicos, Aristide Mingozzi, and Paolo Toth. 1981. State-space relaxation procedures for the computation of bounds to routing problems. *Networks* 11 (2): 145–164.

Chvátal, Vašek, William Cook, George B. Dantzig, Delbert R. Fulkerson, and Selmer M. Johnson. 2010. Solution of a large-scale traveling-salesman problem. *50 Years of Integer Programming 1958–2008*, 7–28. Berlin: Springer.

Clausen, Clifford, and Harry Wechsler. 2000. Quad-q-learning. *IEEE Transactions on Neural Networks* 11 (2): 279–294.

Cleophas, Catherine, and Jan F. Ehmke. 2014. When are deliveries profitable? *Business and Information Systems Engineering* 6 (3): 153–163.

Coelho, Leandro C., Jean-François Cordeau, and Gilbert Laporte. 2014. Heuristics for dynamic and stochastic inventory-routing. *Computers and Operations Research* 52: 55–67.

Coelho, Leandro C., Jacques Renaud, and Gilbert Laporte. 2015. Road-based goods transportation: A survey of real-world applications from 2000 to 2015. Technical report, FSA-2015-007, Québec, Canada.

Côté, Jean-François, Michel Gendreau, and Jean-Yves Potvin. 2013. The vehicle routing problem with stochastic two-dimensional items. Technical report, CIRRELT.

Crainic, Teodor G. 2000. Service network design in freight transportation. *European Journal of Operational Research* 122 (2): 272–288.

Dablanc, Laetitia. 2007. Goods transport in large european cities: Difficult to organize, difficult to modernize. *Transportation Research Part A: Policy and Practice* 41 (3): 280–285.

Delage, Erwann. 2010. Re-optimization of technician tours in dynamic environments with stochastic service time. Technical report, Ecole des Mines de Nantes.

Delen, Dursun. 2014. *Real-World Data Mining: Applied Business Analytics and Decision Making*. Upper Saddle River: FT Press.

Dennis, William T. 2011. *Parcel and Small Package Delivery Industry*. Createspace.

DHL. 2013. Dhl improves flexibility for road freight services through mobile solution. www.dhl.com/content/g0/en/press/releases/releases_2013/logistics/dhl_improves_flexibility_for_road_freight_services_through_mobile_solution.html. Accessed 25 Aug 2015.

Duin, Cees, and Stefan Voß. 1999. The pilot method: A strategy for heuristic repetition with application to the steiner problem in graphs. *Networks* 34 (3): 181–191.

Dyer, James S., and Rakesh K. Sarin. 1982. Relative risk aversion. *Management Science* 28 (8): 875–886.

Economist Intelligence Unit. 2013a. Forecast for major categories regarding u.s. healthcare spending in 2015 and 2020 (in million u.s. dollars). *Economist*. http://www.statista.com/statistics/380534/forecast-for-major-categories-in-us-healthcare-spending/. Accessed 22 Sept 2015.

Economist Intelligence Unit. 2013b. Fostering a data-driven culture. *The Economist*. https://www.tableau.com/sites/default/files/whitepapers/tableau_dataculture_130219.pdf. Accessed 25 Aug 2015.

Ehmke, Jan F. 2012. *Integration of Information and Optimization Models for Routing in City Logistics*, vol. 177. International Series in Operations Research and Management Science. Berlin: Springer.

Ehmke, Jan F., and Ann M. Campbell. 2014. Customer acceptance mechanisms for home deliveries in metropolitan areas. *European Journal of Operational Research* 233 (1): 193–207.

Ehmke, Jan F., Ann M. Campbell, and Timothy L. Urban. 2015. Ensuring service levels in routing problems with time windows and stochastic travel times. *European Journal of Operational Research* 240 (2): 539–550.

Ehmke, Jan F., Ann M. Campbell, and Barrett W. Thomas. 2016. Data-driven approaches for emissions-minimized paths in urban areas. *Computers and Operations Research* 67: 34–47.

Erera, Alan L., Martin Savelsbergh, and Emrah Uyar. 2009. Fixed routes with backup vehicles for stochastic vehicle routing problems with time constraints. *Networks* 54 (4): 270–283.

Errico, Fausto, G. Desaulniers, M. Gendreau, W. Rei, and L.M. Rousseau. 2013. The vehicle routing problem with hard time windows and stochastic service times. Technical report, Cahier du GERAD.

Errico, F., G. Desaulniers, M. Gendreau, W. Rei, and L.-M. Rousseau. 2016. A priori optimization with recourse for the vehicle routing problem with hard time windows and stochastic service times. *European Journal of Operational Research* 249 (1): 55–66.

Esser, Klaus, Judith Kurte. 2015. Kep 2015. Marktanalyse, Bundesverband Paket und Expresslogistik e. V.

European Automobile Manufacturers' Association. 2015. Number of car sharing users worldwide from 2006 to 2014 (in millions). http://www.statista.com/statistics/415636/car-sharing-number-of-users-worldwide/. Accessed 21 Sept 2015.

Evans, James R., and Carl H. Lindner. 2012. Business analytics: the next frontier for decision sciences. *Decision Line* 43 (2): 4–6.

Feo, Thomas A., and Mauricio G.C. Resende. 1995. Greedy randomized adaptive search procedures. *Journal of Global Optimization* 6 (2): 109–133.

Ferreira, João C, Vítor Monteiro, and João L Afonso. 2013. Dynamic range prediction for an electric vehicle. *Electric Vehicle Symposium and Exhibition 2013*, 1–11. IEEE.

Figliozzi, Miguel A. 2007. Analysis of the efficiency of urban commercial vehicle tours: Data collection, methodology, and policy implications. *Transportation Research Part B: Methodological* 41 (9): 1014–1032.

Finkel, Raphael A., and Jon L. Bentley. 1974. Quad trees: a data structure for retrieval on composite keys. *Acta Informatica* 4 (1): 1–9.

Flatberg, Truls, Geir Hasle, Oddvar Kloster, Eivind J. Nilssen, and Atle Riise. 2007. Dynamic and stochastic vehicle routing in practice. *Dynamic Fleet Management*, 41–63. Berlin: Springer.

Fraunhofer. 2015. Marktvolumen des logistikmarktes in europa in den jahren 2008 bis 2014 (in milliarden euro). http://de.statista.com/statistik/daten/studie/204132/umfrage/volumen-des-logistikmarktes-in-europa/. Accessed 21 Nov 2015.

Freedman, David A. 2009. *Statistical Models: Theory and Practice*. Cambridge: Cambridge University Press.

Frigg, Roman, and Stephan Hartmann. 2012. Models in science. *The Stanford Encyclopedia of Philosophy*. The Metaphysics Research Lab (online available).

Garey, Michael R., and David S. Johnson. 1979. *A Guide to the Theory of NP-Completeness*. New York: WH Freemann.

Gendreau, Michel, and Jean-Yves Potvin. 1998. *Dynamic Vehicle Routing and Dispatching*. Berlin: Springer.

Gendreau, Michel, Francois Guertin, Jean-Yves Potvin, and Eric Taillard. 1999. Parallel tabu search for real-time vehicle routing and dispatching. *Transportation Science* 33 (4): 381–390.

Gendreau, Michel, Francois Guertin, Jean-Yves Potvin, and René Séguin. 2006. Neighborhood search heuristics for a dynamic vehicle dispatching problem with pick-ups and deliveries. *Transportation Research Part C: Emerging Technologies* 14 (3): 157–174.

George, Abraham, and Warren B. Powell. 2007. An adaptive-learning framework for semi-cooperative multi-agent coordination. Technical report, Princeton University.

George, Abraham, Warren B. Powell, Sanjeev R. Kulkarni, and Sridhar Mahadevan. 2008. Value function approximation using multiple aggregation for multiattribute resource management. *Journal of Machine Learning Research* 9 (10): 2079–2111.

Gevaers, Roel, Eddy Van de Voorde, and Thierry Vanelslander. 2011. Characteristics and typology of last-mile logistics from an innovation perspective in an urban context. *City Distribution and Urban Freight Transport: Multiples Perspectives*, 56–71.

Ghannadpour, Seyed F., Siamak Noori, and Reza Tavakkoli-Moghaddam. 2013. Multiobjective dynamic vehicle routing problem with fuzzy travel times and customers' satisfaction in supply chain management. *IEEE Transactions on Engineering Management* 60 (4): 777–790.

Ghiani, Gianpaolo, Francesca Guerriero, Gilbert Laporte, and Roberto Musmanno. 2003. Real-time vehicle routing: Solution concepts, algorithms and parallel computing strategies. *European Journal of Operational Research* 151 (1): 1–11.

Ghiani, Gianpaolo, Emanuele Manni, Antonella Quaranta, and Chefi Triki. 2009. Anticipatory algorithms for same-day courier dispatching. *Transportation Research Part E: Logistics and Transportation Review* 45 (1): 96–106.

Ghiani, Gianpaolo, Emanuele Manni, and Barrett W. Thomas. 2012. A comparison of anticipatory algorithms for the dynamic and stochastic traveling salesman problem. *Transportation Science* 46 (3): 374–387.

Golden, Bruce L., Larry Levy, and Rakesh Vohra. 1987. The orienteering problem. *Naval Research Logistics (NRL)* 34 (3): 307–318.

Gómez, Andrés, Ricardo Mariño, Raha Akhavan-Tabatabaei, Andrés L. Medaglia, Jorge E. Mendoza. 2015. On modeling stochastic travel and service times in vehicle routing. *Transportation Science* (online available).

Goodson, Justin C., Jeffrey W. Ohlmann, and Barrett W. Thomas. 2013a. Rollout policies for dynamic solutions to the multivehicle routing problem with stochastic demand and duration limits. *Operations Research* 61 (1): 138–154.

Goodson, Justin C., Barrett W. Thomas, and Jeffrey W. Ohlmann. 2013b. Restocking-based rollout policies for the vehicle routing problem with stochastic demand and duration limits. *Transportation Science* (online available).

Goodson, Justin C., Barrett W. Thomas, and Jeffrey W. Ohlmann. 2014. A generalized rollout algorithm framework for stochastic dynamic programming. Technical report, University of Iowa.

Gouveia, Luis, and Stefan Voß. 1995. A classification of formulations for the (time-dependent) traveling salesman problem. *European Journal of Operational Research* 83 (1): 69–82.

Groß, Patrick-Oliver, Marlin W. Ulmer, and Dirk C. Mattfeld. 2015. Exploiting travel time information for reliable routing in city logistics. *Transportation Research Procedia* 10: 652–661.

Guerriero, Francesca, and Marco Mancini. 2003. A cooperative parallel rollout algorithm for the sequential ordering problem. *Parallel Computing* 29 (5): 663–677.

Hadjiconstantinou, E., and D. Roberts. 2001. Routing under uncertainty: An application in the scheduling of field service engineers. In *The Vehicle Routing Problem*, ed. Paolo Toth, and Daniele Vigo, 331–352. Philadelphia: Society for Industrial and Applied Mathematics.

Haight, Frank A. 1967. *Handbook of the Poisson Distribution*. New York: Wiley.

Häll, Carl H., Jan T. Lundgren, and Stefan Voß. 2012. Evaluating the performance of a dial-a-ride service using simulation. *Public Transport* 1–19.

Hamagami, Tomoki, and Hironori Hirata. 2005. State space partitioning and clustering with sensor alignment for autonomous robots. In *2005 IEEE International Conference on Systems, Man and Cybernetics*, vol. 3, 2655–2660. IEEE.

Han, Jinil, Chungmok Lee, and Sungsoo Park. 2013. A robust scenario approach for the vehicle routing problem with uncertain travel times. *Transportation Science* 48 (3): 373–390.

Hand, David J., Heikki Mannila, and Padhraic Smyth. 2001. *Principles of data mining*. Cambridge: MIT Press.

Hansen, Pierre, Vittorio Maniezzo, and Stefan Voß. 2009. Special issue on mathematical contributions to metaheuristics editorial. *Journal of Heuristics* 15 (3): 197.

Hendrickx, Iris, and Antal Van Den Bosch. 2005. Hybrid algorithms with instance-based classification. *Machine Learning: ECML 2005*, 158–169. Berlin: Springer.

Huet, Ellen, and Liyan Chen. 2015. World war uber: Why the ride-hailing giant can't conquer the planet (yet). *Forbes*. http://www.forbes.com/sites/ellenhuet/2015/09/07/world-war-uber-why-the-ride-hailing-company-cant-conquer-the-planet-yet/. Accessed 12 Sept 2015.

Hvattum, Lars M., Arne Løkketangen, and Gilbert Laporte. 2006. Solving a dynamic and stochastic vehicle routing problem with a sample scenario hedging heuristic. *Transportation Science* 40 (4): 421–438.

Hvattum, Lars M., Arne Løkketangen, and Gilbert Laporte. 2007. A branch-and-regret heuristic for stochastic and dynamic vehicle routing problems. *Networks* 49 (4): 330–340.

Ichoua, Soumia, Michel Gendreau, and Jean-Yves Potvin. 2000. Diversion issues in real-time vehicle dispatching. *Transportation Science* 34 (4): 426–438.

Ichoua, Soumia, Michel Gendreau, and Jean-Yves Potvin. 2006. Exploiting knowledge about future demands for real-time vehicle dispatching. *Transportation Science* 40 (2): 211–225.

Irvine, Chris. 2013. Beijing to impose odd-even car ban during heavy pollution. *The Telegraph.* http://www.telegraph.co.uk/news/worldnews/asia/china/10386738/Beijing-to-impose-odd-even-car-ban-during-heavy-pollution.html. Accessed 22 Sept 2015.

Jaana, Remes, Smit Sven, Manyika James, Woetzel Jonathan, and Agyenim-Boateng Yaw. 2013. Urban world: The shifting global business landscape. Technical report, McKinsey Global Institute, USA.

Jaillet, Patrick. 1988. A priori solution of a traveling salesman problem in which a random subset of the customers are visited. *Operations Research* 36 (6): 929–936.

Jiang, Daniel R., and Warren B. Powell. 2015. Approximate dynamic programming for dynamic quantile-based risk measures. Technical report, Princeton University.

Jula, Hossein, Maged Dessouky, Petros Ioannou, et al. 2006. Truck route planning in nonstationary stochastic networks with time windows at customer locations. *IEEE Transactions on Intelligent Transportation Systems* 7 (1): 51–62.

Kall, Peter, and Stein W. Wallace. 1994. *Stochastic Programming*. New York: Wiley.

Kelley, John L. 1975. *General Topology*. Berlin: Springer Science & Business Media.

Kelly, Heather. 2012. Police embracing tech that predicts crimes. http://edition.cnn.com/2012/07/09/tech/innovation/police-tech/1. Accessed 22 Sept 2015.

Kenyon, Astrid S., and David P. Morton. 2003. Stochastic vehicle routing with random travel times. *Transportation Science* 37 (1): 69–82.

Kim, Seong-Hee, and Barry L. Nelson. 2001. A fully sequential procedure for indifference-zone selection in simulation. *ACM Transactions on Modeling and Computer Simulation (TOMACS)* 11 (3): 251–273.

Kim, Seong-Hee, and Barry L. Nelson. 2006. On the asymptotic validity of fully sequential selection procedures for steady-state simulation. *Operations Research* 54 (3): 475–488.

Kopalle, Praveen. 2014. Why amazon's anticipatory shipping is pure genius. *Forbes*. http://www.forbes.com/sites/onmarketing/2014/01/28/why-amazons-anticipatory-shipping-is-pure-genius/. Accessed 27 Aug 2015.

Köster, Felix, Marlin W. Ulmer, and Dirk C. Mattfeld. 2015. Cooperative traffic control management for city logistic routing. *Transportation Research Procedia* 10: 673–682.

Lahyani, Rahma, Mahdi Khemakhem, and Frédéric Semet. 2012. Taxonomy for rich vehicle routing problems. In *Proceedings of the 1st International Conference on Logistics Operations Management, IEEE International Conference-GOL'12* (online available).

Lahyani, Rahma, Mahdi Khemakhem, and Frédéric Semet. 2015. Rich vehicle routing problems: From a taxonomy to a definition. *European Journal of Operational Research* 241 (1): 1–14.

Land, Ailsa H., and Alison G. Doig. 1960. An automatic method of solving discrete programming problems. *Econometrica: Journal of the Econometric Society* 497–520.

Laporte, Gilbert, Francois Louveaux, and Hélène Mercure. 1992. The vehicle routing problem with stochastic travel times. *Transportation Science* 26 (3): 161–170.

Larsen, Allan, Oli B.G. Madsen, and Marius M. Solomon. 2002. Partially dynamic vehicle routing-models and algorithms. *Journal of the Operational Research Society* 53 (6): 637–646.

Larsen, Allan, Oli B.G. Madsen, and Marius M. Solomon. 2008. Recent developments in dynamic vehicle routing systems. *The Vehicle Routing Problem: Latest Advances and New Challenges*, 199–218. Berlin: Springer.

Lecluyse, Christophe, Tom Van Woensel, and Herbert Peremans. 2009. Vehicle routing with stochastic time-dependent travel times. *4OR* 7 (4): 363–377.

Lee, Ivan S.K., and Henry Y.K. Lau. 2004. Adaptive state space partitioning for reinforcement learning. *Engineering Applications of Artificial Intelligence* 17 (6): 577–588.

Leukel, Joerg, Stefan Kirn, and Thomas Schlegel. 2011. Supply chain as a service: a cloud perspective on supply chain systems. *IEEE Systems Journal* 5 (1): 16–27.

Li, Xiangyong, Peng Tian, and Stephen C.H. Leung. 2010. Vehicle routing problems with time windows and stochastic travel and service times: Models and algorithm. *International Journal of Production Economics* 125 (1): 137–145.
Lin, Erwin T.J., Lawrence W. Lan, and Cathy S.T. Hsu. 2010. Assessing the on-road route efficiency for an air-express courier. *Journal of Advanced Transportation* 44 (4): 256–266.
Lin, Jenn-Rong, and Ta-Hui Yang. 2011. Strategic design of public bicycle sharing systems with service level constraints. *Transportation Research Part E: Logistics and Transportation Review* 47 (2): 284–294.
Lorini, Sandro, Jean-Yves Potvin, and Nicolas Zufferey. 2011. Online vehicle routing and scheduling with dynamic travel times. *Computers and Operations Research* 38 (7): 1086–1090.
Lowe, John, Atif A. Khan, and Bhakti Bhatale. 2014. Same-day delivery: Surviving and thriving in a world where instant gratification rules. Whitepaper 20, Cognizant.
Malandraki, Chryssi, and Mark S. Daskin. 1992. Time dependent vehicle routing problems: Formulations, properties and heuristic algorithms. *Transportation Science* 26 (3): 185–200.
Malandraki, Chryssi, and Robert B. Dial. 1996. A restricted dynamic programming heuristic algorithm for the time dependent traveling salesman problem. *European Journal of Operational Research* 90 (1): 45–55.
Mangalindan, J.P. 2013. The race for same-day delivery. *Fortune*. http://fortune.com/2013/12/05/the-race-for-same-day-delivery/. Accessed 25 Aug 2015.
Manyika, J., M. Chui, B. Brown, J. Bughin, R. Dobbs, C. Roxburgh, and A.H. Byers. 2011. Big data: The next frontier for innovation, competition, and productivity. McKinsey Global Institute, USA.
Maxwell, Matthew S., Mateo Restrepo, Shane G. Henderson, and Huseyin Topaloglu. 2010. Approximate dynamic programming for ambulance redeployment. *INFORMS Journal on Computing* 22 (2): 266–281.
Meisel, Stephan. 2011. *Anticipatory Optimization for Dynamic Decision Making*, vol. 51. Operations Research/Computer Science Interfaces Series. Berlin: Springer.
Meisel, Stephan, and Dirk C. Mattfeld. 2010. Synergies of operations research and data mining. *European Journal of Operational Research* 206 (1): 1–10.
Meisel, Stephan, Uli Suppa, and Dirk C. Mattfeld. 2009. Grasp based approximate dynamic programming for dynamic routing of a vehicle. Technical report, Technische Universität Braunschweig, Germany.
Meisel, Stephan, Uli Suppa, and Dirk C. Mattfeld. 2011. Serving multiple urban areas with stochastic customer requests. *Dynamics in Logistics*, 59–68. Berlin: Springer.
Mes, Martijn, Matthieu van der Heijden, and Peter Schuur. 2010. Look-ahead strategies for dynamic pickup and delivery problems. *OR Spectrum* 32 (2): 395–421.
Mitrović-Minić, Snežana, and Gilbert Laporte. 2004. Waiting strategies for the dynamic pickup and delivery problem with time windows. *Transportation Research Part B: Methodological* 38 (7): 635–655.
Montemanni, Roberto, and Luca M. Gambardella. 2004. An exact algorithm for the robust shortest path problem with interval data. *Computers and Operations Research* 31 (10): 1667–1680.
Montemanni, Roberto, János Barta, Monaldo Mastrolilli, and Luca M. Gambardella. 2007. The robust traveling salesman problem with interval data. *Transportation Science* 41 (3): 366–381.
Mu, Qianxin, Zhuo Fu, Jens Lysgaard, and R. Eglese. 2011. Disruption management of the vehicle routing problem with vehicle breakdown. *Journal of the Operational Research Society* 62 (4): 742–749.
Nemhauser, George L., and Laurence A. Wolsey. 1988. *Integer and Combinatorial Optimization*. Interscience Series in Discrete Mathematics and Optimization. New York: Wiley.
Nemhauser, George L., Alexander H.G. Rinnooy Kan, and Michael J. Kan. 1989. *Optimization, Handbooks in Operations Research and Management Science*, vol. 1. Amsterdam: North-Holland.
Novoa, Clara, and Robert Storer. 2009. An approximate dynamic programming approach for the vehicle routing problem with stochastic demands. *European Journal of Operational Research* 196 (2): 509–515.

Organization for Economic Cooperation and Development. 2013. Delivering the goods: 21st century challenges to urban goods transport. Technical report, OECD Publishing.

Pandelis, Dimitrios G., E.G. Kyriakidis, and T.D. Dimitrakos. 2012. Single vehicle routing problems with a predefined customer sequence, compartmentalized load and stochastic demands. *European Journal of Operational Research* 217 (2): 324–332.

Papastavrou, Jason D. 1996. A stochastic and dynamic routing policy using branching processes with state dependent immigration. *European Journal of Operational Research* 95 (1): 167–177.

Pillac, Victor, Michel Gendreau, Christelle Guéret, and Andrés L. Medaglia. 2013. A review of dynamic vehicle routing problems. *European Journal of Operational Research* 225 (1): 1–11.

Polya, George. 2014. *How to solve it: A new aspect of mathematical method*. Princeton: Princeton University Press.

Powell, Warren B. 2007. *Approximate Dynamic Programming: Solving the Curses of Dimensionality*, vol. 703. Wiley Series in Probability and Statistics. New York: Wiley.

Powell, Warren B. 2009. What you should know about approximate dynamic programming. *Naval Research Logistics (NRL)* 56 (3): 239–249.

Powell, Warren B. 2011. *Approximate Dynamic Programming: Solving the Curses of Dimensionality*, vol. 842. Wiley Series in Probability and Statistics. New York: Wiley.

Powell, Warren B. 2014. Clearing the jungle of stochastic optimization. *Bridging Data and Decisions*, chap. 5. INFORMS, 109–137.

Powell, W.B., and S. Meisel. 2016. Tutorial on stochastic optimization in energy, part I: Modeling and policies. *IEEE Transactions on Power Systems* 31 (2): 1459–1467.

Powell, Warren B., and Ilya O. Ryzhov. 2012. *Optimal Learning*, vol. 841. Wiley Series in Probability and Statistics. New York: Wiley.

Powell, Warren B., Yosef Sheffi, Kenneth S. Nickerson, Kevin Butterbaugh, and Susan Atherton. 1988. Maximizing profits for north american van lines' truckload division: A new framework for pricing and operations. *Interfaces* 18 (1): 21–41.

Powell, Warren B., Michael T. Towns, and Arun Marar. 2000. On the value of optimal myopic solutions for dynamic routing and scheduling problems in the presence of user noncompliance. *Transportation Science* 34 (1): 67–85.

Powell, Warren B., Hugo P. Simao, and Belgacem Bouzaiene-Ayari. 2012. Approximate dynamic programming in transportation and logistics: a unified framework. *EURO Journal on Transportation and Logistics* 1 (3): 237–284.

Psaraftis, Harilaos N. 1980. A dynamic programming solution to the single vehicle many-to-many immediate request dial-a-ride problem. *Transportation Science* 14 (2): 130–154.

Psaraftis, Harilaos N. 1988. Dynamic vehicle routing problems. *Vehicle Routing: Methods and Studies* 16: 223–248.

Psaraftis, Harilaos N., Min Wen, Christos A. Kontovas. 2015. Dynamic vehicle routing problems: Three decades and counting. *Networks* (online available).

Pureza, Vitória, and Gilbert Laporte. 2008. Waiting and buffering strategies for the dynamic pickup and delivery problem with time windows. *INFOR: Information Systems and Operational Research* 46 (3) 165–176.

Puterman, Martin L. 2014. *Markov decision processes: discrete stochastic dynamic programming*. New York: Wiley.

Quak, H.J., and M. (René) B.M. de Koster. 2009. Delivering goods in urban areas: how to deal with urban policy restrictions and the environment. *Transportation Science* 43 (2): 211–227.

Raviv, Tal, and Ofer Kolka. 2013. Optimal inventory management of a bike-sharing station. *IIE Transactions* 45 (10): 1077–1093.

Respen, Jean, Nicolas Zufferey, and Jean-Yves Potvin. 2014. Impact of online tracking on a vehicle routing problem with dynamic travel times. Technical report, CIRRELT.

Ritzinger, Ulrike, Jakob Puchinger, and Richard F. Hartl. 2014. Dynamic programming based metaheuristics for the dial-a-ride problem. *Annals of Operations Research* 1–18.

Ritzinger, Ulrike, Jakob Puchinger, and Richard F. Hartl. 2015. A survey on dynamic and stochastic vehicle routing problems. *International Journal of Production Research* 1–17.

Rosen, Robert. 2012. *Anticipatory systems*. Berlin: Springer.

Rosenkrantz, Daniel J., Richard E. Stearns, and Philip M. Lewis. 1974. Approximate algorithms for the traveling salesperson problem. In *IEEE Conference Record of 15th Annual Symposium on Switching and Automata Theory*, 33–42. IEEE.

Ross, Sheldon M. 2014. *Introduction to probability models*. New York: Academic Press.

Ruszczyński, Andrzej. 2010. Risk-averse dynamic programming for markov decision processes. *Mathematical Programming* 125 (2): 235–261.

Sáez, Doris, Cristián E. Cortés, and Alfredo Núñez. 2008. Hybrid adaptive predictive control for the multi-vehicle dynamic pick-up and delivery problem based on genetic algorithms and fuzzy clustering. *Computers and Operations Research* 35 (11): 3412–3438.

Salvendy, Gavriel. 2001. *Handbook of industrial engineering: technology and operations management*. New York: Wiley.

Sarasola, Briseida, Karl F. Doerner, Verena Schmid, and Enrique Alba. 2015. Variable neighborhood search for the stochastic and dynamic vehicle routing problem. *Annals of Operations Research* 1–37.

Savelsbergh, Martin, and Marc Sol. 1998. Drive: Dynamic routing of independent vehicles. *Operations Research* 46 (4): 474–490.

Schafer, Andreas, and David G. Victor. 2000. The future mobility of the world population. *Transportation Research Part A: Policy and Practice* 34 (3): 171–205.

Schilde, Michael, Karl F. Doerner, and Richard F. Hartl. 2011. Metaheuristics for the dynamic stochastic dial-a-ride problem with expected return transports. *Computers and Operations Research* 38 (12): 1719–1730.

Schilde, Michael, Karl F. Doerner, and Richard F. Hartl. 2014. Integrating stochastic time-dependent travel speed in solution methods for the dynamic dial-a-ride problem. *European Journal of Operational Research* 238 (1): 18–30.

Schmid, Verena. 2012. Solving the dynamic ambulance relocation and dispatching problem using approximate dynamic programming. *European Journal of Operational Research* 219 (3): 611–621.

Schneeweiss, Christoph. 1999. *Hierarchies in Distributed Decision Making*. Berlin: Springer Science & Business Media.

Secomandi, Nicola. 2000. Comparing neuro-dynamic programming algorithms for the vehicle routing problem with stochastic demands. *Computers and Operations Research* 27 (11): 1201–1225.

Secomandi, Nicola. 2001. A rollout policy for the vehicle routing problem with stochastic demands. *Operations Research* 49 (5): 796–802.

Secomandi, Nicola, and Francois Margot. 2009. Reoptimization approaches for the vehicle-routing problem with stochastic demands. *Operations Research* 57 (1): 214–230.

Sheridan, Patricia K., Erich Gluck, Qi Guan, Thomas Pickles, Barış Balcıog, Beno Benhabib, et al. 2013. The dynamic nearest neighbor policy for the multi-vehicle pick-up and delivery problem. *Transportation Research Part A: Policy and Practice* 49: 178–194.

Sinha, Sanskrity. 2013. Dhl's parcelcopter drone for speedy aerial delivery takes off in germany. *International Business Times*. http://www.ibtimes.co.uk/dhl-parcelcopter-paketkopter-amazon-prime-air-drone-528873. Accessed 25 Aug 2015.

Slater, Alan. 2002. Specification for a dynamic vehicle routing and scheduling system. *International Journal of Transport Management* 1 (1): 29–40.

Soeffker, Ninja, Marlin W. Ulmer, and Dirk C. Mattfeld. 2016. Problem-specific state space partitioning for dynamic vehicle routing problems. In *Proceedings of MKWI*, 229–240.

Solomon, Marius M. 1987. Algorithms for the vehicle routing and scheduling problems with time window constraints. *Operations Research* 35 (2): 254–265.

Souyris, Sebastián, Cristián E. Cortés, Fernando Ordóñez, and Andres Weintraub. 2013. A robust optimization approach to dispatching technicians under stochastic service times. *Optimization Letters* 7 (7): 1549–1568.

Spiegel, Joel R., Michael T. McKenna, Girish S. Lakshman, and Paul G. Nordstrom. 2012. Method and system for anticipatory package shipping. US Patent 8,271,398.

Spivey, Michael Z., and Warren B. Powell. 2004. The dynamic assignment problem. *Transportation Science* 38 (4): 399–419.

Steinsiek, Dennis. 2015. 800 cities, 1 million bicycles in bike sharing worldwide. *European Cyclists' Federation*. http://www.ecf.com/news/800-cities-1-million-bicycles-in-bike-sharing-worldwide/. Accessed 21 Sept 2015.

Stock, James R., and Douglas M. Lambert. 2001. *Strategic logistics management*, vol. 4. Boston: McGraw-Hill/Irwin.

Sungur, Ilgaz, Yingtao Ren, Fernando Ordóñez, Maged Dessouky, and Hongsheng Zhong. 2010. A model and algorithm for the courier delivery problem with uncertainty. *Transportation Science* 44 (2): 193–205.

Susilawati, Susilawati, Michael A.P. Taylor, and Sekhar V.C. Somenahalli. 2013. Distributions of travel time variability on urban roads. *Journal of Advanced Transportation* 47 (8): 720–736.

Swihart, Michael R., and Jason D. Papastavrou. 1999. A stochastic and dynamic model for the single-vehicle pick-up and delivery problem. *European Journal of Operational Research* 114 (3): 447–464.

Tang, Hao, and Elise Miller-Hooks. 2005. Algorithms for a stochastic selective travelling salesperson problem. *Journal of the Operational Research Society* 56 (4): 439–452.

Taniguchi, Eiichi, Russell G. Thompson, Tadashi Yamada, and Ron Van Duin. 2001. *City Logistics. Network modelling and intelligent transport systems*. Emerald Group Publishing Limited.

Tassiulas, Leandros. 1996. Adaptive routing on the plane. *Operations Research* 44 (5): 823–832.

Thomas, Barrett W. 2007. Waiting strategies for anticipating service requests from known customer locations. *Transportation Science* 41 (3): 319–331.

Thomas, Barrett W., and Chelsea C. White III. 2004. Anticipatory route selection. *Transportation Science* 38 (4): 473–487.

Thomas, Barrett W., and Chelsea C. White III. 2007. The dynamic shortest path problem with anticipation. *European Journal of Operational Research* 176 (2): 836–854.

TomTom. 2015a. The ten most traffic jam prone cities in europe (as of 2015; longer travel time compared to uncongested traffic). http://www.statista.com/statistics/235780/most-traffic-jam-prone-cities-in-europe/. Accessed 21 Sept 2015.

TomTom. 2015b. The ten most traffic jam prone cities in north america (as of 2015; longer travel times compared with uncongested traffic). http://www.statista.com/statistics/235786/most-traffic-jam-prone-cities-in-north-america/. Accessed 21 Sept 2015.

Toth, Paolo, and Daniele Vigo. 2001. *The vehicle routing problem*. Philadelphia: Society for Industrial and Applied Mathematics.

Trudeau, Pierre, and Moshe Dror. 1992. Stochastic inventory routing: Route design with stockouts and route failures. *Transportation Science* 26 (3): 171–184.

Ulmer, Marlin W., and Marco Hennig. 2016. Value function approximation-based limited horizon rollout algorithms for dynamic multi-period routing. Submitted.

Ulmer, Marlin W., and Dirk C. Mattfeld. 2013. Modeling customers in the Euclidean plane for routing applications. In *Proceedings of the 14th EU/ME Workshop*, 98–103.

Ulmer, Marlin W., and Stefan Voß. 2016. Risk-averse anticipation for dynamic vehicle routing. To appear in Computational Logistics.

Ulmer, Marlin W., Jan Brinkmann, and Dirk C. Mattfeld. 2015a. Anticipatory planning for courier, express and parcel services. *Logistics Management*, 313–324. Berlin: Springer.

Ulmer, Marlin W., Dirk C. Mattfeld, Marco Hennig, and Justin C. Goodson. 2015b. A rollout algorithm for vehicle routing with stochastic customer requests. *Logistics Management*, 217–227. Berlin: Springer.

Ulmer, Marlin W., Justin C. Goodson, Dirk C. Mattfeld, and Barrett W. Thomas. 2016a. Route-based Markov decision processes for dynamic vehicle routing problems. Submitted.

Ulmer, Marlin W., Dirk C. Mattfeld, and Ninja Soeffker. 2016b. Dynamic multi-period vehicle routing: approximate value iteration based on dynamic lookup tables. Technical report, Technische Universität Braunschweig, Germany.

Ulmer, Marlin W., Justin C. Goodson, Dirk C. Mattfeld, and Marco Hennig. 2017a. Offline-online approximate dynamic programming for dynamic vehicle routing with stochastic requests. To appear in Transportation Science.

Ulmer, Marlin W., Dirk C. Mattfeld, and Felix Köster. 2017b. Budgeting time for dynamic vehicle routing with stochastic customer requests. To appear in Transportation Science.

United Nations. 2010. Development of the global population aged 65 and over between 2010 and 2050, by region. http://www.statista.com/statistics/279795/development-of-the-global-population-aged-65-and-over/. Accessed 22 Sept 2015.

United Nations. 2015. Proportion of population in cities worldwide from 1985 to 2050. http://www.statista.com/statistics/264651/proportion-of-population-in-cities-worldwide/. Accessed 21 Sept 2015.

U.S. Energy Information Administration. 2015. U.s. carbon dioxide emissions from energy consumption between 1950 and 2014, by end-use sector (in million metric tons of carbon dioxide). http://www.statista.com/statistics/204879/us-carbon-dioxide-emissions-by-sector-since-1950/. Accessed 21 Sept 2015.

Van Hemert, Jano I., and Johannes A. La Poutré. 2004. Dynamic routing problems with fruitful regions: Models and evolutionary computation. *Parallel Problem Solving from Nature-PPSN VIII*, 692–701. Berlin: Springer.

Van Roy, Benjamin, Dimitri P. Bertsekas, Yuchun Lee, and John N. Tsitsiklis. 1997. A neuro-dynamic programming approach to retailer inventory management. In *Proceedings of the 36th IEEE Conference on Decision and Control*, vol. 4, 4052–4057. IEEE.

Van Woensel, Tom, Laoucine Kerbache, Herbert Peremans, and Nico Vandaele. 2008. Vehicle routing with dynamic travel times: A queueing approach. *European Journal of Operational Research* 186 (3): 990–1007.

Verma, Amit, and Ann M. Campbell. 2014. Strategic placement of telemetry to reduce routing costs. *Networks* 63 (3): 260–275.

Vinod, Hrishikesh D. 1969. Integer programming and the theory of grouping. *Journal of the American Statistical Association* 64 (326): 506–519.

Voccia, Stacy A., Ann Melissa Campbell, and Barrett W. Thomas. 2017. The same-day delivery problem for online purchases. To appear in Transportation Science.

Vogel, Patrick, and Dirk C. Mattfeld. 2011. Strategic and operational planning of bike-sharing systems by data mining–a case study. *Computational Logistics*, 127–141. Berlin: Springer.

Vogel, Patrick, Torsten Greiser, and Dirk C. Mattfeld. 2011. Understanding bike-sharing systems using data mining: Exploring activity patterns. *Procedia-Social and Behavioral Sciences* 20: 514–523.

Wahba, Phil. 2015. Amazon waives same-day delivery fees for prime members. *Fortune*. http://fortune.com/2015/05/28/amazon-same-day-delivery-prime/. Accessed 27 Aug 2015.

Wahl, R., M. Flø, T. Moen, B. Johannessen, T. Engen, and T. Haugen. 2002. Elektronisk datafangst for godstransport i byområder. Technical report, SINTEF.

Xiang, Zhihai, Chengbin Chu, and Haoxun Chen. 2008. The study of a dynamic dial-a-ride problem under time-dependent and stochastic environments. *European Journal of Operational Research* 185 (2): 534–551.

Yan, Shangyao, Jenn-Rong Lin, and Chun-Wei Lai. 2013. The planning and real-time adjustment of courier routing and scheduling under stochastic travel times and demands. *Transportation Research Part E: Logistics and Transportation Review* 53: 34–48.

Yuan, Biao, Ran Liu, and Zhibin Jiang. 2015. A branch-and-price algorithm for the home health care scheduling and routing problem with stochastic service times and skill requirements. *International Journal of Production Research* (online available).

Zadeh, Lotfi A. 1965. Fuzzy sets. *Information and Control* 8 (3): 338–353.

Zhang, Shu, Jeffrey W. Ohlmann, and Barrett W. Thomas. 2014. A priori orienteering with time windows and stochastic wait times at customers. *European Journal of Operational Research* 239 (1): 70–79.

Zhang, Shu, Jeffrey W. Ohlmann, and Barrett W. Thomas. 2015a. Dynamic orienteering on a network of queues. Technical report, University of Iowa.

Zhang, Zhenzhen, Mengyang Liu, and Andrew Lim. 2015b. A memetic algorithm for the patient transportation problem. *Omega* 54: 60–71.

Zielinski, Pawel. 2004. The computational complexity of the relative robust shortest path problem with interval data. *European Journal of Operational Research* 158 (3): 570–576.

Index

© Springer International Publishing AG 2017

M.W. Ulmer, *Approximate Dynamic Programming for Dynamic Vehicle Routing*, Operations Research/Computer Science Interfaces Series 61,
DOI 10.1007/978-3-319-55511-9

MIX
Papier aus verantwortungsvollen Quellen
Paper from responsible sources
FSC® C105338

If you have any concerns about our products,
you can contact us on
ProductSafety@springernature.com

In case Publisher is established outside the EU,
the EU authorized representative is:
Springer Nature Customer Service Center GmbH
Europaplatz 3, 69115 Heidelberg, Germany

Printed by Libri Plureos GmbH
in Hamburg, Germany